Phone Phreaking

Contents

1 Phreaking   **1**
- 1.1 History 1
  - 1.1.1 Switch hook and tone dialer 2
  - 1.1.2 2600 hertz 2
  - 1.1.3 Multi frequency 2
  - 1.1.4 Blue boxes 3
  - 1.1.5 Computer hacking 4
  - 1.1.6 Toll fraud 4
  - 1.1.7 Diverters 5
  - 1.1.8 Voice mail boxes and bridges 5
  - 1.1.9 Cell phones 5
  - 1.1.10 End of multi-frequency 6
- 1.2 2600 Hz 6
- 1.3 In popular culture 7
- 1.4 See also 7
- 1.5 References 8
- 1.6 External links 8

2 2600 hertz   **10**
- 2.1 References 10
- 2.2 See also 10

3 Mark Abene   **12**
- 3.1 Early life 12
- 3.2 Legal tribulations 12
- 3.3 Social protests 13
- 3.4 Professional life 13
- 3.5 References 14
  - 3.5.1 Bibliography 14
  - 3.5.2 Notes 14

| | 3.6 | External links | 15 |

4 BBS: The Documentary — 16
- 4.1 Episodes . . . 16
  - 4.1.1 Disc 1 . . . 16
  - 4.1.2 Disc 2 . . . 16
  - 4.1.3 Disc 3 . . . 16
- 4.2 References . . . 17
- 4.3 External links . . . 17

5 BlueBEEP — 18
- 5.1 References . . . 19
- 5.2 External links . . . 19

6 Direct Access Test Unit — 20
- 6.1 References . . . 20

7 John Draper — 21
- 7.1 Background . . . 21
- 7.2 Phreaking . . . 21
- 7.3 Software developer . . . 23
- 7.4 Health and lifestyle . . . 24
- 7.5 Legends . . . 24
- 7.6 In popular culture . . . 26
- 7.7 See also . . . 26
- 7.8 References . . . 26
- 7.9 External links . . . 27

8 The Hacker Crackdown — 28
- 8.1 Historical perspective . . . 28
- 8.2 Critical reception . . . 28
- 8.3 Quotations . . . 28
- 8.4 References . . . 29
- 8.5 External links . . . 29

9 Joybubbles — 30
- 9.1 Whistler . . . 30
- 9.2 Presence on the media . . . 31
- 9.3 Phone services . . . 31
- 9.4 References . . . 31
- 9.5 External links . . . 32

10 Patrick K. Kroupa — 33

- 10.1 Early years 33
- 10.2 The MindVox Years 34
  - 10.2.1 Voices in my Head (1991–1996) 34
  - 10.2.2 MIA / DOA (1996–2000) 34
- 10.3 21st century 35
  - 10.3.1 Yippies and the counterculture 35
- 10.4 Bibliography 36
  - 10.4.1 Essays 36
  - 10.4.2 Magazines 36
  - 10.4.3 Medical journals 36
- 10.5 References 36
  - 10.5.1 Books 36
  - 10.5.2 Magazines and newspapers 37
  - 10.5.3 Medical journals 37
  - 10.5.4 Public Access U.S. government documents 37
  - 10.5.5 Film 37
  - 10.5.6 Television 37
  - 10.5.7 Radio 37
  - 10.5.8 Music 38
- 10.6 Notes 38
- 10.7 External links 39

11 Elias Ladopoulos — 40

- 11.1 Founding of MOD 40
  - 11.1.1 Conflict with former Legion of Doom members 40
  - 11.1.2 Prosecution 40
- 11.2 Career 41
- 11.3 References 41
- 11.4 External links 42

12 Lucky225 — 43

- 12.1 See also 43
- 12.2 Notable Links 43

13 Phone hacking — 44

- 13.1 Risks 44
- 13.2 Techniques 44
  - 13.2.1 Voicemail 44

	13.2.2 Handsets	46
	13.2.3 Other	46
13.3	Legality	46
13.4	See also	46
13.5	References	47
13.6	External links	48

14 Phantom Access — 49

14.1	History	49
14.2	Hacking as art	50
14.3	Phantom Access 5.7K	50
14.4	Final Evolution	51
14.5	External links	51
14.6	References	51

15 Phone Losers of America — 52

15.1	History	52
	15.1.1 The e-zine	52
	15.1.2 PLA Radio	53
	15.1.3 PLA TV	53
	15.1.4 PLA Voice Bridge	53
	15.1.5 PLA Community Forums	53
	15.1.6 PLA: The Books	54
	15.1.7 Tenth Anniversary	54
15.2	Cactus	55
15.3	Press	55
15.4	United Phone Losers; a PLA Spin-Off	55
15.5	The Snow Plow Show	56
15.6	See also	57
15.7	References	57
15.8	External links	57

16 Phreaking boxes — 59

16.1	List of phreaking boxes	59
16.2	See also	59
16.3	External links	60

17 Plover-NET — 61

17.1	Naming and Creation	61
17.2	Legion of Doom	61

17.3 2600	62

- 17.3 2600 62
- 17.4 Operations 62
- 17.5 References 62

18 The Secret History of Hacking — 63

- 18.1 See also 63
- 18.2 References 63
- 18.3 External links 63

19 StankDawg — 64

- 19.1 Biography 64
- 19.2 Hacking 64
- 19.3 Digital DawgPound 65
  - 19.3.1 History 65
  - 19.3.2 Members 65
  - 19.3.3 Recognition 66
- 19.4 Binary Revolution 66
  - 19.4.1 Binary Revolution Radio 66
  - 19.4.2 BinRev Meetings 67
  - 19.4.3 BinRev.net 67
  - 19.4.4 HackTV 67
  - 19.4.5 DocDroppers 68
- 19.5 Selected writing 68
- 19.6 Selected presentations 68
- 19.7 Projects 68
- 19.8 References 69
- 19.9 External links 69

20 Dennis Terry — 71

- 20.1 References 71

21 ToneLoc — 72

- 21.1 See also 72
- 21.2 References 72
- 21.3 External links 72

22 Van Eck phreaking — 73

- 22.1 Basic principle 73
  - 22.1.1 CRTs 73
  - 22.1.2 LCDs 74

22.1.3 Communicating using Van Eck phreaking	74
22.1.4 Tailored Access Batteries	74
22.2 Countermeasures	74
22.3 See also	74
22.4 References	75
22.5 External links	75

23 War dialing — 76

- 23.1 Process . . . 76
- 23.2 Popularity . . . 76
- 23.3 Variants . . . 77
- 23.4 See also . . . 77
- 23.5 References . . . 77
- 23.6 External links . . . 77

24 WarVOX — 78

- 24.1 See also . . . 78
- 24.2 References . . . 78
- 24.3 External links . . . 78

25 Matthew Weigman — 79

- 25.1 Early life . . . 79
- 25.2 First offense . . . 79
- 25.3 Learning to hack telephones . . . 79
- 25.4 Current incarceration . . . 80
- 25.5 See also . . . 80
- 25.6 References . . . 80
- 25.7 External links . . . 81
- 25.8 Text and image sources, contributors, and licenses . . . 82
  - 25.8.1 Text . . . 82
  - 25.8.2 Images . . . 84
  - 25.8.3 Content license . . . 86

Chapter 1

Phreaking

This article is about the manipulation of telephone call routing. For the use of telephone technology to steal information, see Phone hacking.

Phreaking is a slang term coined to describe the activity of a culture of people who study, experiment with, or explore, telecommunication systems, such as equipment and systems connected to public telephone networks. The term *phreak* is a sensational spelling of the word *freak* with the *ph-* from *phone*, and may also refer to the use of various audio frequencies to manipulate a phone system. *Phreak*, *phreaker*, or *phone phreak* are names used for and by individuals who participate in phreaking.

The term first referred to groups who had reverse engineered the system of tones used to route long-distance calls. By recreating these tones, phreaks could switch calls from the phone handset, allowing free calls to be made around the world. To ease the creation of these tones, electronic tone generators known as blue boxes became a staple of the phreaker community, including future Apple Inc. cofounders Steve Jobs and Steve Wozniak.

The blue box era came to an end with the ever increasing use of computerized phone systems, which sent dialing information on a separate, inaccessible channel. By the 1980s, much of the system in the US and Western Europe had been converted. Phreaking has since become closely linked with computer hacking.[1] This is sometimes called the H/P culture (with *H* standing for *hacking* and *P* standing for *phreaking*).

1.1 History

Phone phreaking got its start in the late 1950s in the United States. Its golden age was the late 1960s and early 1970s. Phone phreaks spent a lot of time dialing around the telephone network to understand how the phone system worked, engaging in activities such as listening to the pattern of tones to figure out how calls were routed, reading obscure telephone company technical journals, learning how to impersonate operators and other telephone company personnel, digging through telephone company trash bins to find "secret" documents, sneaking into telephone company buildings at night and wiring up their own telephones, building electronic devices called blue boxes, black boxes, and red boxes to help them explore the network and make free phone calls, hanging out on early conference call circuits and "loop arounds" to communicate with one another and writing their own newsletters to spread information.

Prior to 1984, long-distance telephone calls were a premium item, with archaic regulations. In some locations, calling across the street counted as long distance.[2] To report that a phone call was long distance meant an elevated importance universally accepted because the calling party is paying by the minute to speak to the called party; transact business quickly.

Phreaking consisted of techniques to evade the long-distance charges. This evasion was illegal; the crime was called "toll fraud."[3]

1.1.1 Switch hook and tone dialer

Possibly one of the first phreaking methods was switch-hooking, which allows placing calls from a phone where the rotary dial or keypad has been disabled by a key lock or other means to prevent unauthorized calls from that phone. It is done by rapidly pressing and releasing the switch hook to open and close the subscriber circuit, simulating the pulses generated by the rotary dial. Even most current telephone exchanges support this method, as they need to be backward compatible with old subscriber hardware.[4]

By rapidly clicking the hook for a variable number of times at roughly 5 to 10 clicks per second, separated by intervals of roughly one second, the caller can dial numbers as if they were using the rotary dial. The pulse counter in the exchange counts the pulses or clicks and interprets them in two possible ways. Depending on continent and country, one click with a following interval can be either "one" or "zero" and subsequent clicks before the interval are additively counted. This renders ten consecutive clicks being either "zero" or "nine", respectively. Some exchanges allow using additional clicks for special controls, but numbers 0–9 now fall in one of these two standards. One special code, "flash", is a very short single click, possible but hard to simulate. Back in the day of rotary dial, very often technically identical phone sets were marketed in multiple areas of the world, only with plugs matched by country and the dials being bezeled with the local standard numbers.

Such key-locked telephones, if wired to a modern DTMF capable exchange, can also be exploited by a tone dialer that generates the DTMF tones used by modern keypad units. These signals are now very uniformly standardized worldwide, and along with rotary dialing, they are almost all that is left of in-band signaling. It is notable that the two methods can be combined: Even if the exchange does not support DTMF, the key lock can be circumvented by switch-hooking, and the tone dialer can be then used to operate automated DTMF controlled services that can't be used with rotary dial.

1.1.2 2600 hertz

Main article: 2600 hertz

The origins of phone phreaking trace back at least to AT&T's implementation of fully automatic switches. These switches used tone dialing, a form of in-band signaling, and included some tones which were for internal telephone company use. One internal-use tone was a tone of 2600 Hz which caused a telephone switch to think the call had ended, leaving an open carrier line, which could be exploited to provide free long-distance, and international, calls. At that time, long-distance calls were quite expensive.[5]

The tone was discovered in approximately 1957,[5] by Joe Engressia, a blind seven-year-old boy. Engressia had perfect pitch, and discovered that whistling the fourth E above middle C (a frequency of 2600 Hz) would stop a dialed phone recording. Unaware of what he had done, Engressia called the phone-company and asked why the recordings had stopped. Joe Engressia is considered to be the father of phreaking.[6]

Other early phreaks, such as "Bill from New York" (William "Bill" Acker 1953-2015), began to develop a rudimentary understanding of how phone networks worked. Bill discovered that a recorder he owned could also play the tone at 2600 Hz with the same effect. John Draper discovered through his friendship with Engressia that the free whistles given out in Cap'n Crunch cereal boxes also produced a 2600 Hz tone when blown (providing his nickname, "Captain Crunch"). This allowed control of phone systems that worked on single frequency (SF) controls. One could sound a long whistle to reset the line, followed by groups of whistles (a short tone for a "1", two for a "2", etc.) to dial numbers.[7]

1.1.3 Multi frequency

Main article: Multi-frequency

While single-frequency worked on certain phone routes, the most common signaling on the then long-distance network was multi-frequency (MF) controls. The slang term for these tones and their use was "Marty Freeman." The specific frequencies required were unknown to the general public until 1964, when the Bell System published the information in the *Bell System Technical Journal* in an article describing the methods and frequencies used for inter-office signalling.

The journal was intended for the company's engineers; however, it found its way to various college campuses across the United States. With this one article, the Bell System accidentally gave away the "keys to the kingdom," and the intricacies of the phone system were at the disposal of people with a knowledge of electronics.[8]

The second generation of phreaks arose at this time, including the New Yorkers "Evan Doorbell", "Ben Decibel" and Neil R. Bell and Californians Mark Bernay, Chris Bernay, and "Alan from Canada". Each conducted their own independent exploration and experimentation of the telephone network, initially on an individual basis, and later within groups as they discovered each other in their travels. "Evan Doorbell," "Ben" and "Neil" formed a group of phreaks, known as Group Bell. Mark Bernay initiated a similar group named the Mark Bernay Society. Both Mark and Evan received fame amongst today's phone phreakers for Internet publication of their collection of telephone exploration recordings. These recordings, conducted in the 1960s, 1970s, and early 1980s are available at Mark's website *Phone Trips*.[9]

1.1.4 Blue boxes

Blue Box

Main article: Blue box

In October 1971, phreaking was introduced to the masses when Esquire Magazine published a story called "Secrets of the Little Blue Box"[10][11][12][13] by Ron Rosenbaum. This article featured Engressia and John Draper prominently, synonymising their names with phreaking. The article also attracted the interest of other soon-to-be phreaks, such as Steve Wozniak and Steve Jobs, who went on to found Apple Computer.[14][15]

1971 also saw the beginnings of YIPL (Youth International Party Line), a publication started by Abbie Hoffman and Al Bell to provide information to Yippies on how to "beat the man," mostly involving telephones. In the first issue of YIPL, writers included a "shout-out" to all of the phreakers who provided technological information for the newsletter: "We at YIPL would like to offer thanks to all you phreaks out there."[16] At the end of the issue, YIPL stated:

> YIPL believes that education alone cannot affect the System, but education can be an invaluable tool for those willing to use it. Specifically, YIPL will show you why something must be done immediately in regard,

of course, to the improper control of the communication in this country by none other than bell telephone company.[16]

In 1973, Al Bell would move YIPL over and start TAP (Technological American Party).[17] TAP would develop into a major source for subversive technical information among phreaks and hackers all over the world. TAP ran from 1973 to 1984, with Al Bell handing over the magazine to "Tom Edison" in the late 70s. TAP ended publication in 1984 due mostly to a break-in and arson at Tom Edison's residence in 1983.[18] Cheshire Catalyst then took over running the magazine for its final (1984) year.

A controversially suppressed article "How to Build a 'Phone Phreaks' box" in Ramparts Magazine (June, 1972) touched off a firestorm of interest in phreaking. This article published simple schematic plans of a "black box" used to make free long-distance phone calls, and included a very short parts list that could be used to construct one. Bell sued Ramparts, forcing the magazine to pull all copies from shelves, but not before numerous copies were sold and many regular subscribers received them.

1.1.5 Computer hacking

In the 1980s, the revolution of the personal computer and the popularity of computer bulletin board systems (BBSes) (accessed via modem) created an influx of tech-savvy users. These BBSes became popular for computer hackers and others interested in the technology, and served as a medium for previously scattered independent phone phreaks to share their discoveries and experiments. This not only led to unprecedented collaboration between phone phreaks, but also spread the notion of phreaking to others who took it upon themselves to study, experiment with, or exploit the telephone system. This was also at a time when the telephone company was a popular subject of discussion in the US, as the monopoly of AT&T Corporation was forced into divestiture. During this time, exploration of telephone networks diminished, and phreaking focused more on toll fraud. Computer hackers began to use phreaking methods to find the telephone numbers for modems belonging to businesses, which they could exploit later. Groups then formed around the BBS hacker/phreaking (H/P) community such as the famous Masters of Deception (Phiber Optik) and Legion of Doom (Erik Bloodaxe) groups. In 1985, an underground e-zine called Phrack (a combination of the words Phreak and Hack) began circulation among BBSes, and focused on hacking, phreaking, and other related technological subjects.

In the early 1990s, H/P groups like Masters of Deception and Legion of Doom were shut down by the US Secret Service's Operation Sundevil. Phreaking as a subculture saw a brief dispersion in fear of criminal prosecution in the 1990s, before the popularity of the internet initiated a reemergence of phreaking as a subculture in the US and spread phreaking to international levels.

Into the turn of the 21st century, phreaks began to focus on the exploration and playing with the network, and the concept of toll fraud became widely frowned on among serious phreakers, primarily under the influence of the website Phone Trips, put up by second generation phreaks Mark Bernay and Evan Doorbell.

1.1.6 Toll fraud

The 1984 AT&T breakup gave rise to many small companies intent upon competing in the long distance market. These included the then-fledgling Sprint and MCI, both of whom had only recently entered the marketplace. At the time, there was no way to switch a phone line to have calls automatically carried by non-AT&T companies. Customers of these small long distance operations would be required to dial a local access number, enter their calling card number, and finally enter the area code and phone number they wish to call. Because of the relatively lengthy process for customers to complete a call, the companies kept the calling card numbers short – usually 6 or 7 digits. This opened up a huge vulnerability to phone phreaks with a computer.

6-digit calling card numbers only offer 1 million combinations. 7-digit numbers offer just 10 million. If a company had 10,000 customers, a person attempting to "guess" a card number would have a good chance of doing so correctly once every 100 tries for a 6-digit card and once every 1000 tries for a 7-digit card. While this is almost easy enough for people to do manually, computers made the task far easier.[19][20] "Code hack" programs were developed for computers with modems. The modems would dial the long distance access number, enter a random calling card number (of the proper number of digits), and attempt to complete a call to a computer bulletin board system (BBS). If the computer connected successfully to the BBS, it proved that it had found a working card number, and it saved that number to disk. If it did not connect to the BBS in a specified amount of time (usually 30 or 60 seconds), it would hang up and try a different code. Using this method, code hacking programs would turn up hundreds (or in some cases thousands) of working calling card numbers per day. These would subsequently be shared amongst fellow phreakers.

There was no way for these small phone companies to identify the culprits of these hacks. They had no access to local phone company records of calls into their access numbers, and even if they had access, obtaining such records would be prohibitively expensive and time-consuming. While there was some advancement in tracking down these code hackers in the early 1990s, the problem did not completely disappear until most long distance companies were able to offer standard 1+ dialing without the use of an access number.

1.1.7 Diverters

Another method of obtaining free phone calls involved the use of so-called "diverters". Call forwarding was not an available feature for many business phone lines in the 1980s and early 1990s, so they were forced to buy equipment that could do the job manually between two phone lines. When the business would close, they would program the call diverting equipment to answer all calls, pick up another phone line, call their answering service, and bridge the two lines together. This gave the appearance to the caller that they were directly forwarded to the company's answering service. The switching equipment would typically reset the line after the call had hung up and timed out back to dial tone, so the caller could simply wait after the answering service had disconnected, and would eventually get a usable dial tone from the second line. Phreakers recognized the opportunity this provided, and they would spend hours manually dialing businesses after hours, attempting to identify faulty diverters. Once a phreaker had access to one of these lines, he could use it for one of many purposes. In addition to completing phone calls anywhere in the world at the businesses' expense, they could also dial 1-900 phone sex/entertainment numbers, as well as use the phone line to harass their enemies without fear of being traced. Victimized small businesses were usually required to foot the bill for the long distance calls, as it was their own private equipment (not phone company security flaws) that allowed such fraud to occur. By 1993, call forwarding was offered to nearly every business line subscriber, making these diverters obsolete. As a result, hackers stopped searching for the few remaining ones, and this method of toll fraud died.

1.1.8 Voice mail boxes and bridges

Prior to the BBS era of the 1980s phone phreaking was more of a solitary venture as it was difficult for phreaks to connect with one another. In addition to communicating over BBSs phone phreaks discovered voice mail boxes and party lines as ways to network and keep in touch over the telephone. It was rare for a phone phreak to legally purchase access to voice mail. Instead, they would usually appropriate unused boxes that were part of business or cellular phone systems. Once a vulnerable mailbox system was discovered, word would spread around the phreak community, and scores of them would take residence on the system. They would use the system as a "home base" for communication with one another until the rightful owners would discover the intrusion and wipe them off. Voice mailboxes also provided a safe phone number for phreaks to give out to one another as home phone numbers would allow the phreak's identity (and home address) to be discovered. This was especially important given that phone phreaks were breaking the law.

Phreakers also used "bridges" to communicate live with one another. The term "bridge" originally referred to a group of telephone company test lines that were bridged together giving the effect of a party-line. Eventually, all party-lines, whether bridges or not, came to be known as bridges if primarily populated by hackers and/or phreakers.

The popularity of the Internet in the mid-1990s, along with the better awareness of voice mail by business and cell phone owners, made the practice of stealing voice mailboxes less popular. To this day bridges are still very popular with phreakers yet, with the advent of VoIP, the use of telephone company owned bridges has decreased slightly in favor of phreaker-owned conferences.

1.1.9 Cell phones

By the late 1990s, the fraudulent aspect of phreaking all but vanished. Most cellular phones offered unlimited domestic long distance calling for the price of standard airtime (often totally unlimited on weekends), and flat-rate long-distance plans appeared offering unlimited home phone long distance for as little as $25 per month. Rates for international calls had also decreased significantly. Between the much higher risk of being caught (due to advances in technology) and the much lower gain of making free phone calls, toll fraud started to become a concept associated very little with phreaking.

1.1.10 End of multi-frequency

The end of multi-frequency (MF) phreaking in the lower 48 United States occurred on June 15, 2006, when the last exchange in the contiguous United States to use a "phreakable" MF-signalled trunk replaced the aging (yet still well kept) N2 carrier with a T1 carrier. This exchange, located in Wawina Township, Minnesota, was run by the Northern Telephone Company of Minnesota.

1.2 2600 Hz

In the original analog networks, short-distance telephone calls were completed by sending relatively high-power electrical signals through the wires to the end office, which then switched the call. This technique could not be used for long-distance connections, because the signals would be filtered out due to capacitance in the wires. Long-distance switching remained a manual operation years after short-distance calls were automated, requiring operators at either end of the line to set up the connections.

Bell automated this process by sending "in-band" signals. Since the one thing the long-distance trunks were definitely able to do was send voice-frequency signals, the Bell System used a selection of tones sent over the trunks to control the system. When calling long-distance, the local end-office switch would first route the call to a special switch which would then convert further dialing into tones and send them over an appropriately selected trunk line (selected with the area code). A similar machine at the far end of the trunk would decode the tones back into electrical signals, and the call would complete as normal.

In addition to dialing instructions, the system also included a number of other tones that represented various commands or status. 2600 Hz, the key to early phreaking, was the frequency of the tone sent by the long-distance switch indicating that the user had gone on-hook (hung up the phone). This normally resulted in the remote switch also going on-hook, freeing the trunk for other uses. In order to make free lines easy to find, the 2600 Hz tone was continually played into free trunks. If the tone was sent manually by the local user into the phone line, it would trigger the remote switch to go on-hook, but critically, the *local* switch knew he was still off-hook because that was signaled electrically, not by the tone (which their local switch ignored). The system was now in an inconsistent state, leaving the local user connected to an operational long-distance trunk line. With further experimentation, the *phreaks* learned the rest of the signals needed to dial on the remote switch.

Normally, long-distance calls were billed locally. Since the "trick" required a long distance call to be placed in order to connect to the remote switch, it would be billed as usual. However, there were some types of calls that had either no billing, like calls to directory service, or for which the billing was reversed or billed to another number, like WATS lines (area code 800 numbers). By dialing one of these "toll-free" numbers, the caller was connected to a remote switch as normal, but no billing record was made locally. The caller would then play the 2600 Hz tone into the line to return the remote switch to on-hook, and then use a blue box to dial the number to which they really wanted to connect. The local Bell office would have no record of the call.

As knowledge of phreaking spread, a minor culture emerged from the increasing number of phone phreaks. Sympathetic (or easily social-engineered) telephone company employees were persuaded to reveal the various routing codes to use international satellites and trunk lines. At the time it was felt that there was nothing Bell could do to stop this. Their entire network was based on this system, so changing the system in order to stop the phreakers would require a massive infrastructure upgrade.

In fact, Bell responded fairly quickly, but in a more targeted fashion. Looking on local records for inordinately long calls to directory service or other hints that phreakers were using a particular switch, filters could then be installed to block efforts at that end office. Many phreakers were forced to use pay telephones as the telephone company technicians regularly tracked long-distance toll free calls in an elaborate cat-and-mouse game. AT&T instead turned to the law for help, and a number of phreaks were caught by the government.

Eventually, the phone companies in North America did, in fact, replace all their hardware. They didn't do it to stop the phreaks, but simply as a matter of course while moving to fully digital switching systems. Unlike the crossbar switch, where the switching signals and voice were carried on the same lines, the new systems used separate signaling lines which phreaks could not access. This system is known as Common Channel Interoffice Signaling. Classic phreaking with the 2600 Hz tone continued to work in more remote locations into the 1980s, but was of little use in North America by the 1990s.

The last 2600 Hz-controlled trunk in the contiguous United States was operated by the independent Northern Telephone Company with an N2 Carrier system serving Wawina, Minnesota until June 15, 2006, when it was replaced by T1 carrier.[21] The last 2600 Hz-controlled trunks in North America were located in Livengood, Alaska, survived another 5 years, and were finally retired in March 2011.[22]

1.3 In popular culture

- In the movie *The Core*, the hacker known as "Rat" uses phreaking to give Doctor Josh Keyes unlimited long distance for life to prove his skills.

- In the *Person of Interest* episode *Aletheia*, main character Harold Finch uses phreaking to call someone in Paris in 1979 as a prank. On October 27, 1980, Finch uses phreaking to hack ARPANET, an action that causes him to be wanted for treason.

1.4 See also

- 2600: The Hacker Quarterly

- BlueBEEP

- Busy line interrupt

- Hacking

- In-Band Signalling

- Novation CAT

- Phil Lapsley, author of *Exploding the Phone*, a comprehensive history of phreaking

- Phone hacking

- Telephone tapping

- Phone Losers of America

- Phrack

- Phreaking boxes

- Social engineering (security)

- Software cracking

1.5 References

[1] Sterling, Bruce (2002) [1993]. *The Hacker Crackdown*. McLean, Virginia: IndyPublish.com. ISBN 1-4043-0641-2.

[2] Stott, Kim (22 July 1983). "Hung Up Glenpool Has Long-Distance Woes In Making Calls Across the Street". *NewsOK*. Retrieved 26 May 2013.

[3] "Notice to our customers regarding Toll Fraud" (PDF). BizFon. Retrieved 2014-07-25.

[4] SoftCab. "Phone Call Recorder". Modemspy.com. Retrieved 2014-07-24.

[5] Robson, Gary D. (April 2004). "The Origins of Phreaking". Blacklisted! 411.

[6] DELON (February 27, 2008). "COMPLETE HISTORY OF HACKING". *Hacking | LEMNISCATE*. Retrieved 2014-12-25.

[7] Lapsley, Phil (2013-11-02). *Exploding the Phone: The Untold Story of the Teenagers and Outlaws who Hacked Ma Bell*. New York: Grove/Atlantic, corporated. ISBN 080212061X.

[8] *Bell System Technical Journal*. **43** (5). September 1964 https://web.archive.org/web/20120314023659/http://www.alcatel-lucent.com/bstj/vol43-1964/bstj-vol43-issue05.html. Archived from the original on March 14, 2012. Retrieved 24 June 2011. Missing or empty |title= (help)

[9] "Phone Trips". Retrieved 2008-06-21.

[10] Rosenbaum, Ron (2011-10-07). "The article that inspired Steve Jobs: "Secrets of the Little Blue Box"". Slate.com. Archived from the original on 2011-11-03. Retrieved 2013-11-30.

[11] "Secrets of the Little Blue Box". Retrieved 2010-09-04.

[12] "Steve Jobs and Me: He said my 1971 article inspired him. His iBook obsessed me.". Retrieved 2011-10-12.

[13] ""Secrets of the Little Blue Box": The 1971 article about phone hacking that inspired Steve Jobs.". Archived from the original on 2011-11-03. Retrieved 2011-10-12.

[14] "Welcome to Woz.org". Retrieved 2008-06-21.

[15] Lapsley, Phil (20 February 2013). "The Definitive Story of Steve Wozniak, Steve Jobs, and Phone Phreaking". theatlantic.com: *The Atlantic*. Archived from the original on 23 February 2013. Retrieved 24 September 2015.

[16] Coleman, Gabriella. *Phreaks, Hackers, and Trolls*. p. 104.

[17] "Youth International Party Line (YIPL) / Technological American Party (TAP), New York FBI files 100-NY-179649 and 117-NY-2905 (3.2 Mbytes)." (PDF). Retrieved 2013-11-30.

[18] "Cheshire's Book - TAP.HTML". Retrieved 2008-06-21.

[19] "W32.Bugbear.B Worm Identified As Targeting Banks | Scoop News". Scoop.co.nz. 2003-06-09. Retrieved 2014-07-24.

[20] Angela Moscaritolo (2011-03-18). "AT&T sues two over scheme to steal customer data". SC Magazine. Retrieved 2014-07-24.

[21] "Telephone World - Sounds & Recordings from Wawina, MN". Phworld.org. Retrieved 2013-11-30.

[22] "The death of Livengood - Old Skool Phreaking - Binary Revolution Forums". Binrev.com. Retrieved 2013-11-30.

1.6 External links

- "Original Esquire article that started it all".
- AusPhreak - Australia's oldest and largest phreaking forum
- Secrets of the Little Blue Box – article with photos
- Telephone World – Sounds & Recordings of Wawina, Minnesota

1.6. EXTERNAL LINKS

- Textfiles.com / phreak Large collection of phreaking related text files. See also, audio conferences.
- Digital Information Society
- The History of Phone Phreaking
- Phone Trips Large collection of historical phone recordings.
- Phreaky Boys A collection of recordings made in 1990 of voice mail box systems compromised by phreakers.
- Phone Phreaking Demonstrated in India.

Chapter 2

2600 hertz

2600 Hz is a frequency in hertz (cycles per second) that was used by AT&T as a steady signal to mark currently unused long-distance telephone lines.

During the 1960s, in-band signaling was used, so the same line for both voice conversations and telephone connection management signals. Since a pause in a voice conversation would produce silence, another method was required for switches to determine available circuits. The solution AT&T created was to produce a 2600 Hz tone on idling trunks.

A device, known as a "blue box", was created to generate the 2600 Hz signal on a line being used. This indicated to switch that the line was idle. After the tone, the switch believed another call was starting, and used the subsequent dialed digits to connect the call.[1]

This technique only affected interoffice multi-frequency (MF) trunks; local calls originated and terminated on the same switch. By placing a call to a non-local toll-free number, interoffice trunks were used for free. Using a blue box would then disconnect the toll-free call and let any other number be dialed. Since the phone was never physically hung up, the connection was still toll-free.

At one point in the 1960s, packets of the Cap'n Crunch breakfast cereal included a free gift: a small whistle that (by coincidence) generated a 2600 Hz tone when one of the whistle's two holes was covered. The phreaker Captain Crunch adopted his nickname from this whistle. Others would utilize exotic birds such as canaries which are able to hit the 2600 Hz tone to the same effect.

In the 1970s and 80s some trunks were modified to filter out SF tone arriving from a caller. Later in the 20th century, long-distance companies adopted the out-of-band signaling system Signaling System 7. This system separated the voice and signaling channels, making it impossible to generate these signals from an ordinary phone line.

2.1 References

[1] Sterling, Bruce. "2". *The Hacker Crackdown.*

2.2 See also

- Phreaking, the general term for exploiting the telephone system in unintended ways
- Falsing
- Red box (phreaking)
- Black box (phreaking)
- Blue box (phreaking)

2.2. SEE ALSO

- Single-frequency signaling
- *2600: The Hacker Quarterly*, a magazine named after the 2600 Hz tone.

Chapter 3

Mark Abene

Mark Abene (born 1972), is an infosec expert and entrepreneur, originally from New York City. Better known by his pseudonym **Phiber Optik**, he was once a member of the hacker groups Legion of Doom and Masters of Deception.

Phiber Optik was a high-profile hacker in the 1980s and early 1990s, appearing in *The New York Times*, *Harper's*, *Esquire*, in debates and on television. He is an important figure in the 1995 non-fiction book *Masters of Deception — The Gang that Ruled Cyberspace*.[1]

3.1 Early life

Mark Abene's first contact with computers was at around 9 years of age at a local department store, where he would often pass the time while his parents shopped. His first computer was a TRS-80 MC-10 with 4 kilobytes of RAM, a 32-column screen, no lower case, and a cassette tape recorder to load and save programs. As was customary at the time, the computer connected to a television set for use as a monitor. After receiving the gifts of a RAM upgrade (to 20K) and a 300 baud modem from his parents, he used his computer to access CompuServe and shortly after discovered the world of dialup BBSes via people he met on CompuServe's "CB simulator", the first nationwide online chat. On some of these BBSes, Abene discovered dialups and guest accounts to DEC minicomputers running the RSTS/E and TOPS-10 operating systems as part of the BOCES educational program in Long Island, New York. Accessing those DEC minicomputers he realized there was a programming environment that was much more powerful than that of his own home computer, and so he began taking books out of the library in order to learn the programming languages that were now available to him. This and the ability to remotely save and load back programs that would still be there the next time he logged in had a profound effect on Abene, who came to view his rather simple computer as a window into a much larger world.[2]

Having learned about programming and fundamental security concepts during those early years, Abene further honed his skill in understanding the intricacies of the nationwide telephone network.

In the mid-1980s he was first introduced to members of the Legion of Doom (LOD), a loosely knit group of highly respected teenage hackers who shared Abene's uncompromising desire to understand technology. Their main focus was to explore telecommunications systems, minicomputer and mainframe operating systems and large-scale packet data networks. The eventual decline of the LOD toward the late 1980s largely due to fragmentation and dissention within the group, coupled with the legal prosecution of a handful of its members, caused Abene to increasingly align himself with a local group of up-and-coming hackers, who came to be known as the Masters of Deception (MOD).

3.2 Legal tribulations

On January 24, 1990, Abene and other MOD members had their homes searched and property seized by the U.S. Secret Service largely based on government suspicions of having caused AT&T Corporation's network crash just over a week earlier on January 15 (Abene was personally accused by the Secret Service of having done as much, during the search and seizure). Some weeks later, AT&T themselves admitted that the crash was the result of a flawed software update to the switching systems on their long distance network, thus, human error on their part.[3] In February 1991, Abene was arrested and charged with computer tampering and computer trespass in the first degree, New York state offenses. Laws at the time were considered a "gray area" concerning information security. Abene, who was a minor at the time, pleaded "not guilty" to the first two offenses and ultimately accepted a plea agreement to a lesser misdemeanor charge, and was sentenced to 35 hours of community service.[4]

Abene and four other members of the Masters of Deception were also arrested in December 1991 and indicted by a Manhattan federal grand jury on July 8, 1992, on an 11-count charge. The indictment relied heavily on evidence collected by court-approved wire tapping of telephone conversations between MOD members. According to U.S. Attorney Otto Obermaier,[4] it was the "first investigative use of court-authorized wiretaps to obtain conversations and data transmissions of computer hackers" in the United States.

According to a July 9, 1992 newsletter from the Electronic Frontier Foundation, the defendants faced a maximum term of 50 years in prison and fines of $2.5 million if found guilty on all counts. Despite the fact that Abene was a minor at the time the crimes were allegedly committed, was only involved in a small fraction of the sub-charges, and often in a passive way, a plea arrangement resulted in by far the harshest sentence: 12 months imprisonment, three years probation and 600 hours of community service.

After serving the one-year sentence at the Federal Prison "Camp" in Schuylkill, Pennsylvania, Abene was released in November 1994. In January 1995, a huge celebration called "Phiberphest '95" was held in his honor at Manhattan's Irving Plaza ballroom/nightclub. In *Time*, Joshua Quittner called him "the first underground hero of the Information Age, the Robin Hood of cyberspace."[5]

3.3 Social protests

Many people inside and outside of the hacker world felt that Abene was made an example of, and was not judged according to earlier court standards. Abene had built up a significant reputation in the hacker sub-culture, for example regularly appearing on the radio show *Off the Hook*, hosted by Eric Corley (a.k.a. Emmanuel Goldstein), debating and defending the morals and motivations of hackers in public forums and in interviews, and lecturing on the history of telecommunications technology at the night courses of several New York City universities. At the time of the indictment he was working at MindVox, an early BBS/ISP founded by two New York LOD members, and subsequently on EchoNYC, a multi-user BBS and early ISP.

ECHO users, ECHO management themselves and hackers around the nation expected Abene to get off with probation or at most a few months of jail time. Co-defendants and previous offenders charged with "hacking" offenses had received rather lenient punishments, and given his new-found enthusiasm for using his knowledge to constructive ends, the general feeling was optimistic prior to sentencing.

A statement made by U.S. Attorney Otto Obermeier in conjunction with the indication, "The message that ought to be delivered with this indictment is that such conduct will not be tolerated, irrespective of the age of the particular accused or their ostensible purpose,"[6] was interpreted by Abene's supporters to mean that MOD was made an example of, to show that the authorities could handle the perceived "hacker threat". During sentencing, Judge Stanton said that "the defendant stands as a symbol here today," and that "hacking crimes constitute a real threat to the expanding information highway", reinforcing the view that a relatively harmless "teacher" was judged as a symbol for all hackers.[7]

3.4 Professional life

Abene has spoken on the subject of security in many publications such as the New York Times, the Washington Post, the Wall Street Journal, and Time Magazine. He has appeared as a speaker at both hacker and security industry conferences worldwide and frequently visits universities to speak to students about information security.[8]

After some years as a security consultant, he joined forces with former Legion of Doom member Dave Buchwald and a third colleague, Andrew Brown, to create the security consulting firm Crossbar Security. Crossbar provided consulting services for third party companies, during which the principals conducted business in the U.S., Japan, Brazil, and Sweden. As a result of the "dot com" bust Crossbar ultimately went defunct in 2001, largely due to cuts in corporate security spending.

Abene made his acting début as "The Inside Man" in the fiction film *Urchin*,[9] completed in 2006 and released in the US in February 2007, in which other hacker notables such as Dave Buchwald and Emmanuel Goldstein can also be seen.

In 2009, he founded TraceVector, an intrusion detection firm that makes use of supercomputing and data analytics. He currently resides in Silicon Valley.

3.5 References

3.5.1 Bibliography

- *The Rise and Fall of Information Security in the Western World*. Speech by Mark Abene, *Hack in the Box* security conference, Kuala Lumpur, Malaysia, 2007.
- CNET Q&A: *Mark Abene, from 'Phiber Optik' to security guru.*
- New York Software Industry Association.
- Goldstein, Emmanuel (2001). Freedom Downtime, opening sequence.
- Savage, Annaliza (September 1995). Notes from the underground — Phiber Optik goes directly to jail. *.net* Issue 10.
- Quittner, Joshua (January 23, 1995). Hacker Homecoming. *TIME*.
- Dibbell, Julian (January 12, 1994). Prisoner: Phiber Optik Goes Directly to Jail. *The Village Voice*
- Sterling, Bruce (January 1994). *The Hacker Crackdown — Law and Disorder on the Electronic Frontier*. . From Project Gutenberg.
- Goldstein, Emmanuel (November 10, 1993). Interview with Phiber Optik. *Off the Hook* radio show. (Online archive)
- Electronic Frontier Foundation (July 9, 1992). Federal hacking indictments issued against five in New York City. Retrieved September 4, 2004
- Newsbytes (July 9, 1992). New York Computer Crime Indictments. Retrieved September 11, 2004.
- Grand jury, United States District Court Southern District of New York (1992). Indictment of Julio Fernandez, John Lee, Mark Abene, Elias Ladopoulos, Paul Stira. (Copy from Computer underground Digest, 4:31).
- *All Circuits are Busy Now: The 1990 AT&T Long Distance Network Collapse.*

3.5.2 Notes

[1] ISBN 978-0-06-092694-6

[2] Q&A: Mark Abene, from 'Phiber Optik' to security guru

[3] All Circuits are Busy Now: The 1990 AT&T Long Distance Network Collapse

[4] Newsbytes

[5] Hacker Homecoming

[6] Newsbytes, 1992

[7] Dibbel, 1994 and Goldstein, 1993, 2001

[8] "The Rise & Fall of Information Security in the Western World" - Keynote, HITB Security Conference, Kuala Lumpur, Malaysia, 2007

[9] http://www.urchinthemovie.com

3.6 External links

- The History of MOD
    - modbook1.txt — *"The History of MOD: Book One: The Originals"*
    - modbook2.txt — *"The History of MOD: Book Two: Creative Mindz"*
    - modbook3.txt — *"The Book of MOD: Part Three: A Kick in the Groin"*
    - modbook4.txt — *"The Book of MOD: Part Four: End of '90-'1991"*
    - modbook5.txt — *"The Book of MOD: Part 5: Who are They And Where Did They Come From? (Summer 1991)"*
- Phiber Optik Goes to Prison — Article in Wired Magazine by Julian Dibbell
- *Off the Hook* shows (available as MP3 files)
    - 1991-03-13, "Phiber Optik's" first appearance on the show. .
    - 1993-11-03, announcement of Mark Abene's sentence. No recording exists. .
    - 1993-11-10, the first show following the sentencing, Phiber Optik in the studio. .
    - 1994-01-05, last show before Phiber Optik's going to prison. .

Chapter 4

BBS: The Documentary

BBS: The Documentary (commonly referred to as ***BBS Documentary***) is a 3-disc, 8-episode documentary about the subculture born from the creation of the bulletin board system (BBS) filmed by computer historian Jason Scott Sadofsky of textfiles.com.

Production work began in July 2001 and completed in December 2004. The finished product began shipping in May 2005.

Although the documentary was released under the Creative Commons Attribute-ShareAlike 2.0 License [1] and later under 3.0,[2] meaning that anyone can legally download it for free, Jason Scott Sadofsky has made it known that the downloadable version is only a taste of the full experience and recommends that individuals purchase the documentary DVDs.

4.1 Episodes

4.1.1 Disc 1

1. **Baud**: the beginnings of the first BBSes, featuring Ward Christensen and Randy Suess [39:12]
2. **SysOps and Users**: experiences from those who used and operated BBSes [44:44]

4.1.2 Disc 2

1. **Make it Pay**: the BBS industry of the 1980s and 90s featuring Philip L. Becker, founder of eSoft [46:48]
2. **FidoNet**: details the largest volunteer-run computer network in history [43:56]
3. **Artscene**: the history of the ANSI Art Scene which thrived in the BBS world [42:41]

4.1.3 Disc 3

1. **HPAC (Hacking Phreaking Anarchy Cracking)**: hear from the users of "underground" BBSes [38:22]
2. **No Carrier**: the end of the dial-up BBS and its integration into the Internet [21:32]
3. **Compression**: the story of the PKWARE/SEA legal battle of the late 1980s [20:46]

Disc 3 also serves as a DVD-ROM which contains thousands of photographs from the 200 interviews recorded during the 4-year production of the film. All of the episodes are subtitled in English and include director's commentary tracks. The Artscene episode is the only one to include subtitles translated into Russian. All discs include hidden easter eggs.

4.2 References

[1] "New BBS documentary released under Creative Commons". Jimgilliam.com. 2005-06-03. Retrieved 2009-08-10.

[2] "BBS: The Documentary". archive.org. 2005. Retrieved 2015-09-05.

4.3 External links

- BBS: The Documentary official website
- *BBS: The Documentary* at the Internet Movie Database
- Some especially interesting scenes from the documentary (from YouTube)
- BBS: The Documentary at the Internet Archive

Chapter 5

BlueBEEP

A happy BlueBEEP user back in 1995

BlueBEEP was a popular blue boxing computer program for MS-DOS written between 1993-1995 by the German programmer Stefan Andreas Scheytt,[1] known by the pseudonym Onkel Dittmeyer. Used correctly, it could be used to exploit vulnerabilities in the CCITT Signaling System No. 5, used by international telephone switches of this era, to make free calls around the world. The program spread like wildfire around the planet via the BBS scene and was popular with phreaks, hackers and the warez community.

The Pascal source code was released to the public along with the final version on April 1, 1995.

BlueBEEP has been praised as "the most finely programmed phreaking tool ever coded".[2]

The install docs report the build system as follows: "386-40 8meg with 530meg HDD and SB/16+SVGA, a Philips 102-key soft-keyboard, a 2001 canadian keyboard and a GENIUS 4-year-old shoplifted 3-button mouse."

5.1 References

[1] BLUEBEEP... E TRISTEZA

[2] Anonymous. "SAMS 'Maximum Security' mentions BlueBEEP". Retrieved 2008-05-23.

5.2 External links

- BlueBEEP Executable
- BlueBEEP v1.00 Pascal Source Code

Chapter 6

Direct Access Test Unit

Direct Access Test Units (**DATUs**) are special PSTN phone numbers that terminate at the central office switch in a telephone company's local exchange that provide switchmen and telco technicians with a circuit for testing lines in various ways.[1]

Among the many things a DATU can do are:

- Make certain electrical connections, such as shorting a line's tip to ring, tip to ground, or ring to ground.

- Play certain tones on the line (high tone, low tone) to test the line's audio circuit.

- Monitor the audio circuit on the line if it is in use (though this only provides a scrambled rendition of the line's audio circuit for privacy and security reasons).

- Create and modify certain variables such as counters and timers on the CO switch, as well as adding and modifying exchanges.

- Remove a line from permanent signal holding state.

DATU's are primarily non-published numbers, though they usually have the phone number XXX-9935 (where XXX represents the local exchange). They are password protected for security, though many telcos leave the default password of 1111 set.

In July 2004, New York phone phreak William Quinn, also known as "decoder", was arrested and charged with accessing Verizon's DATU numbers on over 100 occasions. Verizon subsequently claimed in press releases that it spent $120,000 to change the passwords on all the DATU numbers (it is unclear if this represented Verizon changing the DATU passwords from the default and if so why this had not been done at time of initial installation).[2]

6.1 References

[1] http://www.tech-faq.com/datu.html Tech FAQs

[2] http://www.computerworld.com/securitytopics/security/cybercrime/story/0,10801,94512,00.html Computer World, July 13, 2004

Chapter 7

John Draper

For other people named John Draper, see John Draper (disambiguation).

John Thomas Draper (born 1943), also known as **Captain Crunch**, **Crunch** or **Crunchman** (after the Cap'n Crunch breakfast cereal mascot), is an American computer programmer and former phone phreak. He is a legendary figure within the computer programming world and the hacker and security community. Draper has long maintained a nomadic lifestyle;[1] as of May 2013, he resides in Las Vegas, Nevada.[2]

7.1 Background

Draper is the son of a United States Air Force engineer; he has characterized his father as a distant and imposing figure. As a child, he built a home radio station from discarded military components.[3] He was frequently bullied in school and briefly received psychological treatment due to a perceived "chemical imbalance."[4] After taking college courses, Draper himself entered the Air Force in 1964. While stationed in Alaska, he helped his fellow servicemen make free phone calls home by devising access to a local telephone switchboard. After Alaska, he was stationed at Charleston Air Force Station in Maine. In 1967, he created WKOS [W-"chaos"], a pirate station in nearby Dover-Foxcroft, but had to shut it down when a legitimate radio station, WDME, objected. He was honorably discharged from the Air Force as an airman first class[4] or sergeant[5] in 1968. Thereafter, he relocated to the incipient Silicon Valley and briefly held military-related[3] positions with National Semiconductor (as an engineering technician) and Hugle International (tasked with working on an early cordless telephone design) while concurrently enrolled at De Anza College, where he studied part-time through 1972.[6] During this period, he also worked as an engineer/DJ for KKUP in Cupertino, California[7] and adopted the countercultural ethos of the times, including long hair and a predilection for marijuana.[8] Draper is renowned for his lifelong intolerance of tobacco smoke and his poor personal hygiene.[9][10]

7.2 Phreaking

While Draper was driving around in his Volkswagen Microbus to test a pirate radio transmitter he had built, he broadcast a telephone number to listeners as feedback to gauge his station's reception. A callback from a "Denny" (identified in a Discovery Channel documentary as Denny Teresi[11]) resulted in a meeting that caused him to blunder into the world of the phone phreaks. Teresi and a large percentage of the phone phreaks were blind.[12] Learning of his electronic capability, they wanted him to build a multifrequency tone generator (the "blue box") to gain easier entry into the AT&T system, which was controlled by tones. The group was previously using an organ and cassette recordings of tones to get free calls, although a blind boy who had taken the moniker of Joybubbles had perfect pitch and was able to identify the exact frequencies.

The phreakers informed Draper that a toy whistle that was, at the time, packaged in boxes of Cap'n Crunch cereal could

A Cap'n Crunch *bosun whistle*

emit a tone at precisely 2600 hertz—the same frequency that was used by AT&T long lines to indicate that a trunk line was ready and available to route a new call.[13] This would effectively disconnect one end of the trunk, allowing the still connected side to enter an operator mode. Experimenting with this whistle inspired Draper to build blue boxes: electronic devices capable of reproducing other tones used by the phone company.

> I don't do that. I don't do that anymore at all. And if I do it, I do it for one reason and one reason only. I'm learning about a system. The phone company is a System. A computer is a System, do you understand? If I do what I do, it is only to explore a system. Computers, systems, that's my bag. The phone company is nothing but a computer.
> — *Secrets of the Little Blue Box*, Ron Rosenbaum, *Esquire Magazine* (October 1971)

The class of vulnerabilities Draper and others discovered was limited to call-routing switches that employed in-band signaling, whereas newer equipment relies almost exclusively on out-of-band signaling, the use of separate circuits to transmit voice and signals. Though they no longer serve a practical use, the *Cap'n Crunch* whistles did become valued collector's items. Some hackers sometimes go by the handle "Captain Crunch" even today; *2600: The Hacker Quarterly* is named after this whistle frequency.

The expense of sustaining the unbilled phone calls, the redesign of the line protocols, and the accelerated equipment replacement due to the blue box is difficult to calculate, or even to separate from something as complex and dynamic as the telephone long-distance network.

The ubiquity of the 1971 *Esquire* article precipitated Draper's eventual arrest on toll fraud charges in 1972; he was sentenced to five years' probation. However, it also elicited the attention of University of California, Berkeley engineering student and future Apple co-founder Steve Wozniak, who located Draper while working as an engineer at KKUP, a public radio station in Cupertino, California.[14] After arranging to meet Wozniak in his dorm room (whereupon the latter noted Draper's disheveled appearance and body odor), he began to teach his phone phreaking skills to Wozniak and one of his friends, Steve Jobs; a phone phreaking business formed by Wozniak and Jobs presaged the eventual emergence of Apple.[15] Draper was a member of the Homebrew Computer Club.[13] By 1977, he provided services to Apple as an

Homebrew Computer Club members: John Draper, Lee Felsenstein, Roger Melen.

independent contractor,[7] where he created the "Charlie Board," a telephone interface board for the Apple II personal computer "that could immediately identify phone signals and lines -- such as ones that made free calls -- something modems were not able to do for a decade... the technology would later be used for tone-activated calling menus, voicemail and other purposes."[15] The board was never marketed due to a litany of factors, including the prohibitive cost of a corequisite AT&T-approved connection interface device, industrial suppression by the aforementioned corporation, Draper's previous arrest and conviction for wire fraud, and Jobs' personal animus against Draper.[3] Draper also wrote the BASIC cross-assembler used by Wozniak in the development of Apple I and Apple II.[16]

7.3 Software developer

In 1976 and 1978, Draper served two prison sentences for phone fraud. During this period, "two court-appointed psychiatrists examined... Draper. One concluded he had an 'underdeveloped sense of people' and was 'psychotic'; the other found nothing wrong with him."[15] Draper wrote *EasyWriter*, the first word processor for the Apple II, in 1979 during a third prison sentence (a year of nights, subsequently reduced to several months) in the Alameda County jail.[15] Under a work furlough program, he had access to a computer at Receiving Studios, a small band practice studio, where he coded much of *EasyWriter;* however he did take copies of the code "home" to prison overnight to work on it.[17]

Draper later ported *EasyWriter* to the IBM PC, beating Bill Gates on the bid for the IBM contract; the lucrative deal enabled Draper to buy a Mercedes and a house in Hawaii. However, Draper's company, Capn' Software, posted less than $1 million revenue over six years, and he subsequently sued his software's distributor, Bill Baker, over an unauthorized version of EasyWriter that Baker released without Draper's permission; they eventually settled out of court.

His last major corporate position was with Autodesk in the late 1980s; because of his criminal record and eccentricities, many corporations have demonstrated reluctance to hire Draper in spite of his pedigree and demonstrated qualifications. He spent most of the 1990s enmeshed in the burgeoning rave scene, supporting his itinerant lifestyle by developing websites and writing code in Australia and India, among other locales.[15] A Sydney rave website[18] captures alleged diarised reports from John about some of the rave events he attended in Sydney.

From 1999 to 2004, Draper was the Chief Technical Officer (CTO) for ShopIP,[19] a computer security company whose featured OpenBSD-based firewall, The CrunchBox GE, was backed by Steve Wozniak and featured in an article in *The Register*.[20] Being the first company to feature hackers as security consultants and the first to use OpenBSD, Draper and the company were also featured in an article in *The New York Times*.[21]

From 2005 to 2010, Draper was the Chief Technical Officer (CTO) for media delivery company En2go.[22] He also worked as a senior developer of KanTalk!, a VoIP client.

Draper's software development history includes:

- The Motorola 6800 Cross Assembler for CallComputer (1974)
- The Charlie Board (1977)
- Forth 1.7 for the Apple II (1978)
- EasyWriter© (1980)
- Advanced 3-D Graphic Design Systems for Autodesk(1986–89)
- Website Development (1994 to present)
- The Crunchbox Firewall (CTO of ShopIp 1999–2004)
- VOIP application for OnInstant (2005)
- The Channel Manager for the Flyxo Media System (CTO of En2Go 2005–10)

7.4 Health and lifestyle

Draper is well known in the hacker community for his itinerant and nomadic lifestyle.[1] A 2007 *Wall Street Journal* profile described his one-room apartment in Mountain View, California in a state of squalor, littered with trash and miscellaneous electronics and computers.[3] He has since moved to Las Vegas, Nevada, advertising on Twitter in 2013 that he had room for two people to stay with him during DEFCON.[2]

In 2014, Draper elaborated on some of his health issues, including degenerative spine disease and C. Diff, on his blog. Supporters launched a crowdfunding campaign to aide him with his medical bills, raising over $17,000.[23]

7.5 Legends

In an oft-repeated story, Draper picked up a public phone, then proceeded to "phreak" his call around the world. At no charge, he routed a call through different phone switches in countries such as Japan, Russia and England. Once he had set the call to go through dozens of countries, he dialed the number of the public phone next to him. A few minutes later, the phone next to him rang. Draper spoke into the first phone, and, after quite a few seconds, he heard his own voice very faintly on the other phone. He sometimes repeated this stunt at parties. Draper also claimed that he and a friend once placed a direct call to the White House during the Nixon administration, and after giving the operator President Nixon's secret code name of "Olympus", and asking to speak to the president about a national emergency, they were connected with someone who sounded like Richard Nixon; Draper's friend told the man about a toilet paper shortage in Los Angeles, at which point the person on the other end of the line angrily asked them how they'd managed to get connected to him.[15]

John Draper in Canberra, Australia in 1995

7.6 In popular culture

John Draper appears as himself in the unreleased documentary *Hackers Wanted*.[24]

John Draper's story has also inspired several mentions in popular culture. Elements of the movie *Sneakers* recall Draper and Joybubbles; the character Erwin "Whistler" Emory portrayed by David Strathairn, as well as Cosmo's experience of offering phreaking services to criminals while in prison, were based on them.[25][26]

John Draper is specifically mentioned as "Captain Crunch" in one scene in *Cowboy Bebop: The Movie*, in which a hacker mentions that "Cap'n Crunch broke into the national phone system with a plastic whistle."[27]

He is portrayed by Wayne Pére in the movie *Pirates of Silicon Valley*.[28]

Captain Crunch is being searched for by Rockford during a murder investigation on the TV show *The Rockford Files*, season 5, episode 5, "Kill the Messenger".[29]

In *Ready Player One* by Ernest Cline, John Draper is key to unlocking one of the mysteries within the story.[30]

7.7 See also

- *The Secret History of Hacking*, a 2001 documentary film featuring Draper.

7.8 References

[1] "John Draper Interviewed Early 1995". Barbalet.net. Retrieved 2014-07-17.

[2] 15 May 2013 (2013-05-15). "Twitter / jdcrunchman: @_defcon_ I have room for 2". Twitter.com. Retrieved 2014-07-17.

[3] Rhoads, Chris (2007-01-13). "The Twilight Years of Cap'n Crunch". *Wall Street Journal*. ISSN 0099-9660. Retrieved 2016-05-28.

[4] *Exploding the Phone: The Untold Story of the Teenagers and Outlaws Who ... - Phil Lapsley – Google Books*. Books.google.com. 1972-05-04. Retrieved 2014-07-17.

[5] The release of a revised AFR 39-36 on 19 October 1967 renamed Airman Third Class, Airman Second Class and Airman First Class to Airman, Airman First Class and Sergeant respectively. This returned Sergeant to the rank structure as the first step in the NCO tier as a retention move but required achievement of a 5-skill Air Force Specialty Code (AFSC) level. No changes to the respective insignias were made. Technical Sergeant Spink, Barry L. (1992). *A Chronology of the Enlisted Rank Chevron of the United States Air Force*, 19 February 1992. Air Force Historical Research Agency.

[6] "John Draper". LinkedIn. Retrieved 2014-07-17.

[7] "Stories of Apple – Captain Crunch on Apple – An interview with John Draper". Storiesofapple.net. Retrieved 2014-07-17.

[8] *Exploding the Phone: The Untold Story of the Teenagers and Outlaws Who ... - Phil Lapsley – Google Books*. Books.google.com. Retrieved 2014-07-17.

[9] RHOADS, CHRIS (January 13, 2007). "The Twilight Years of Cap'n Crunch". *The Wall Street Journal*. Retrieved April 16, 2010.

[10] Lapsley, Phil (2013). *Exploding the Phone: The Untold Story of the Teenagers and Outlaws who Hacked Ma Bell*. Grove Press. ISBN 978-0-8021-2061-8.

[11] *TLC Hackers: Computer Outlaws* (AVI), Phreak Vids

[12] "Phone Phreaks: The Fascinating Story of the World's First Hackers – BachelorsDegreeOnline.com". *BachelorsDegreeOnline.com*.

[13] Wozniak, S. G. (2006), *iWoz: From Computer Geek to Cult Icon: How I Invented the Personal Computer, Co-Founded Apple, and Had Fun Doing It*. W. W. Norton & Company. ISBN 0-393-06143-4.

[14] The Woz..., The Real Captain Crunch: Stories, Web Crunchers.

[15] Chris Rhoads (Jan 13, 2007). "The Twilight Years of Cap'n Crunch". The Wall Street Journal. Retrieved Jan 29, 2014.

[16] *Captain Crunch on Apple — An interview with John Draper*, Stories of Apple, 2008-12-04

[17] EasyWriter, Stories, WebCrunchers.

[18] "Sydney Rave History". *Sydney Rave History*.

[19] John Leyden (2001-02-07). "Captain Crunch sets up security firm". theregister.co.uk.

[20] Andrew Orlowski (2002-02-27). "Woz blesses Captain Crunch's new box". theregister.co.uk.

[21] John Markoff (2001-01-29). "The Odyssey Of a Hacker: From Outlaw To Consultant". The New York Times.

[22] Marty Graham (2008-01-15). "Wozniak Backs Captain Crunch in Net Video Startup". Wired (magazine).

[23] "Fans raise cash to help phone phreaker John Draper, aka Cap'n Crunch". *Ars Technica*. Retrieved 2016-05-28.

[24] https://thesprawl.org/simstim/hackers-wanted/

[25] *Sneakers* at the Internet Movie Database

[26] "Sneakers (Film) – TV Tropes". *TV Tropes*.

[27] "Captain Crunch – SOLDIERX.COM". *soldierx.com*.

[28] "Pirates of Silicon Valley (1999) – Full Credits – TCM.com". *Turner Classic Movies*.

[29] "The Rockford Files – Season 5, Episode 5: Kill the Messenger – TV.com". *TV.com*. CBS Interactive.

[30] "Ready Player One". *shmoop.com*.

7.9 External links

- "The Real Captain Crunch", Draper's personal web page.
- "Crunch Creations", Draper's business homepage.
- John Draper interviewed on the TV show Triangulation on the TWiT.tv network
- American Archive of Public Broadcasting: Program about ripping off the phone company

Chapter 8

The Hacker Crackdown

The Hacker Crackdown: Law and Disorder on the Electronic Frontier is a work of nonfiction by Bruce Sterling first published in 1992.

The book discusses watershed events in the hacker subculture in the early 1990s. The most notable topic covered is Operation Sundevil and the events surrounding the 1987-1990 war on the Legion of Doom network: the raid on Steve Jackson Games, the trial of "Knight Lightning" (one of the original journalists of *Phrack*), and the subsequent formation of the Electronic Frontier Foundation. The book also profiles the likes of "Emmanuel Goldstein" (publisher of *2600: The Hacker Quarterly*), the former Assistant Attorney General of Arizona Gail Thackeray, FLETC instructor Carlton Fitzpatrick, Mitch Kapor, and John Perry Barlow.

In 1994, Sterling released the book for the Internet with a new afterword.

8.1 Historical perspective

Though published in 1992, and released as a freeware, electronic book in 1994, the book offers a unique and colorful portrait of the nature of "cyberspace" in the early 1990s, and the nature of "computer crime" at that time. The events that Sterling discusses occur on the cusp of the mass popularity of the Internet, which arguably achieved critical mass in late 1994. It also encapsulates a moment in the information age revolution when "cyberspace" morphed from the realm of telephone modems and BBS' into the Internet and the World Wide Web.

8.2 Critical reception

Cory Doctorow, who voiced an unabridged podcast of the book, said it "inspired me politically, artistically and socially".[1]

8.3 Quotations

> I can see a future in which any person can have a Node on the [Internet]. Any person can be a publisher. It's better than the media we now have. It's possible.
> — Mitch Kapor

> [John Perry] Barlow was the first commentator to adopt William Gibson's striking science-fictional term "cyberspace" as a synonym for the present- day nexus of computer and telecommunications networks. Barlow was insistent that cyberspace should be regarded as a qualitatively new world, a 'frontier.' According to Barlow, the world of electronic communications, now made visible through the computer screen, could no

longer be usefully regarded as just a tangle of high-tech wiring. Instead, it had become a place, cyberspace, which demanded a new set of metaphors, a new set of rules and behaviors. The term, as Barlow employed it, struck a useful chord, and this concept of cyberspace was picked up by *Time*, *Scientific American*, computer police, hackers, and even Constitutional scholars. 'Cyberspace' now seems likely to become a permanent fixture of the language.
— On the initial founding of the Electronic Frontier Foundation

The electronic landscape changes with astounding speed. We are living through the fastest technological transformation in human history. I was glad to have a chance to document cyberspace during one moment in its long mutation; a kind of strobe-flash of the maelstrom.
— From the afterword

8.4 References

[1] Doctorow, Cory (June 23, 2007). "Cory podcasts Bruce Sterling's "The Hacker Crackdown"". Boing Boing.

8.5 External links

- Gutenberg etext of *The Hacker Crackdown*
- Czech translation of *The Hacker Crackdown*
- *The Hacker Crackdown* hosted at MIT
- Audiobook of *The Hacker Crackdown*
- eBooks@Adelaide

(University of Adelaide)

Chapter 9

Joybubbles

Joybubbles (May 25, 1949 – August 8, 2007), born **Josef Carl Engressia, Jr.** in Richmond, Virginia, USA, was an early phone phreak. Born blind, he became interested in telephones at age four.[1] He had absolute pitch, and was able to whistle 2600 hertz into a telephone, an operator tone also used by blue box phreaking devices. Joybubbles said that he had an IQ of "172 or something".[2] Joybubbles died at his Minneapolis home on August 8, 2007 (aged 58). According to his death certificate,[3] he died of "natural causes" with "congestive heart failure" as a contributing condition.

9.1 Whistler

As a five-year-old, Engressia discovered he could dial phone numbers by clicking the hang-up switch rapidly ("tapping"), and at the age of 7 he accidentally discovered that whistling at certain frequencies could activate phone switches.[4]

A student at the University of South Florida in the late 1960s, he was given the nickname "Whistler" due to his ability to place free long-distance phone calls by whistling the proper tones with his mouth. After a Canadian operator reported him for selling such calls for $1 at the university, he was suspended and fined $25 but soon reinstated.[4] He later graduated with a degree in philosophy and moved to Tennessee.

According to FBI records, the phone company SBT&T first noticed his phreaking activities in summer 1968, and an employee of the Florida Bell Telephone Company illegally monitored Engressia's telephone conversations and divulged them to the FBI.[4]

After law enforcement raided his house, he was charged with malicious mischief, given a suspended sentence, and quickly abandoned phreaking.

In 1982, he moved to Minneapolis, Minnesota. He lived off his Social Security disability pension and a job as a test subject for scent-intensity research. He was an ordained minister of his own Church of Eternal Childhood, and ran a one-man nonprofit support organization for people rediscovering and re-experiencing childhood, called "We Won't Grow Up".[5] He tried to remain an active member of the children's community around his home, giving readings at the local library and setting up phone calls to terminally ill children around the world. He often contributed to the Bulletin Board section of the *St. Paul Pioneer Press* newspaper.

Sexually abused as a child by one of his teachers, Joybubbles "reverted to his childhood" in May 1988 and remained there until his death, claiming that he was five years old. He legally changed his name to Joybubbles in 1991, stating that he wanted to put his past, specifically the abuse, behind him.[5] He was listed in the local phone directory as "Joybubbles, I Am".

An avid fan of Mister Rogers, Joybubbles was mentioned in a November 1998 *Esquire* magazine article about children's television host Fred Rogers. In the summer of 1998, Joybubbles traveled to the University of Pittsburgh's *Mister Rogers' Neighborhood* Archives and listened to several hundred episodes over a span of six weeks.[5][6]

An active amateur radio operator with the call sign WB0RPA, he held an amateur extra class license, the highest grade

9.2 Presence on the media

- In 1971, just after his arrest, Engressia was featured in an *Esquire* article by Ron Rosenbaum (*Secrets of the Little Blue Box*) which exposed the phone phreak scene to a general public and led to further media coverage of Engressia, who became a cultural icon.[4]

- The 1992 movie *Sneakers* had a character named "Whistler", who seemed to combine traits of both Joybubbles and John Draper. The character of *Whistler*, played by David Strathairn, is based directly on Engressia.[8]

- The 2001 documentary film *The Secret History of Hacking* features archive footage of Joybubbles.

- In his book *iWoz: From Computer Geek to Cult Icon: How I Invented the Personal Computer, Co-Founded Apple, and Had Fun Doing It*, Apple co-founder Steve Wozniak mentions Joybubbles as an early inspiration during his college years.

- On February 21, 2012, WNYC's *Radiolab* aired a segment on Joybubbles in an episode titled *Escape!*

- Joybubbles is the subject of an upcoming feature-length documentary film.

- Chapter 9 of the book *Exploding the Phone* by Phil Lapsley details his successful plan to get a job by purposely getting arrested for phreaking.[5]

9.3 Phone services

Joybubbles ran a weekly telephone story line called "Stories and Stuff". Stories and Stuff was usually updated on the weekend.

In the early 1980s, he ran a phone line called the "Zzzzyzzerrific Funline", which had the distinction of being the very last entry in the phone book.[5] During the Zzzzyzzerrific Funline days, calling himself Highrise Joe, he would go on various rants about how much he loved Valleyfair amusement park and would also regularly play and discuss Up with People.

9.4 References

[1] "Joe Engressia, Expert 'Phone Phreak,' Dies". *All Things Considered*. National Public Radio. 20 August 2007.

[2] "A Conversation with Joybubbles", *Pittsburgh Post-Gazette*, June 25, 1998, archived from the original on February 18, 2010, retrieved June 10, 2014

[3] "The History of Phone Phreaking Blog: August 27, 2008".

[4] Price, David (30 June 2008). "Blind Whistling Phreaks and the FBI's Historical Reliance on Phone Tap Criminality". CounterPunch.

[5] Lapsley, Phil (2013). *Exploding the Phone: The Untold Story of the Teenagers and Outlaws Who Hacked Ma Bell*. Grove Press. pp. 316–317. ISBN 978-0-8021-9375-9. Retrieved 10 July 2014.

[6] Junod, Tom (November 1998). "Can You Say...Hero?". *Esquire*.

[7] Joybubbles - S.K.

[8] "*Sneakers* Trivia". *IMBD*. Retrieved 10 July 2014.

Previously (context): issued.[7] As shown in the Federal Communications Commission database, he also earned both a General radiotelephone operator license and a commercial radiotelegraph operator's license, as well as a ship radar endorsement on these certificates. He was also one of the few people to qualify for the now-obsolete aircraft radiotelegraph endorsement on the latter license.

9.5 External links

- New York Times Obituary
- New York Times Magazine memorial profile
- Pittsburgh Post-Gazette profile (2003)
- 11-20-91 *Off the Hook* interview / Summary of the *Off The Hook* interview
- An archive of Stories and Stuff
- A Haxor Radio interview with Joybubbles (April 22, 2004)
- Radiolab audio segment describing Joybubbles' background (Feb 2012)
- A video of Joybubbles making a phone call by whistling on YouTube
- A conversation with Joybubbles from 1998
- *Secrets of the Little Blue Box*
- Joybubbles: The Documentary Film
- American Archive of Public Broadcasting: Program about ripping off the phone company

Chapter 10

Patrick K. Kroupa

Patrick Karel Kroupa (also known as **Lord Digital**, born January 20, 1969) is an American writer, hacker and activist. Kroupa was a member of the legendary Legion of Doom and Cult of the Dead Cow hacker groups and co-founded MindVox in 1991, with Bruce Fancher. He was a heroin addict from age 14 to 30 and got clean through the use of the hallucinogenic drug ibogaine.

10.1 Early years

Kroupa was born in Los Angeles, California, of Czech parents who left Prague, Czechoslovakia, after the Soviet invasion in 1968. His parents were divorced when Kroupa was six, and he relocated to New York City, where he was raised by his mother. He is the nephew of Czech opera singer Zdeněk Kroupa.[1]

Patrick Kroupa was part of the first generation to grow up with home computers and network access. In numerous interviews he has repeatedly listed two events which were important in shaping the course of his later years.

The first was being exposed to one of the first two Cray supercomputers that were ever built, which was located at the National Center for Atmospheric Research (NCAR) where his father was a physicist, who took him through the labs and taught him to program in Fortran and feed the Cray using punched cards. This happened during the same year that Woody Allen was filming *Sleeper*, using NCAR in many of the futuristic background scenes that appeared in the movie. Kroupa got an Apple II computer for his own use around the time he was seven or eight years old.[2]

The second event was being part of the last days of Abbie Hoffman's YIPL/TAP (Youth International Party Lines/Technological Assistance Program) counter-culture/Yippie meetings that were taking place in New York City's Lower East Side, during the early 1980s. Kroupa again lists this event, repeatedly in interviews, as opening many new doors for him and changing his perceptions about technology.

TAP was the original hacker and phone phreak publication which predated *2600* by decades (at the time of the last TAP meetings, 2600 magazine was just starting to publish its first issues). Kroupa met many people there who would become part of his life in the years to come. Three of the main characters would be his future partner and lifelong friend, Bruce Fancher; Yippie/Medical Marijuana activist Dana Beal (The Theoretician), who was part of the John Draper (Cap'n Crunch) /Abbie Hoffman, technologically inclined branch of the counter-culture and perhaps most important: Herbert Huncke, who introduced Kroupa to heroin at age 14.[3]

With the exception of the counter-cultural and hard-drug elements, the preceding history made Kroupa part of a small group, composed of a few hundred kids who were either wealthy enough to afford home computers in the late 1970s, or had technologically savvy families who understood the potentials of what the machines could do.[4] The Internet as it is today did not exist; only a small percentage of the population had home computers and out of those who did, even fewer had online access through the use of modems.[5]

During his time in the computer underground Kroupa was a member of the first Pirate/Cracking crew to ever exist for the Apple II computer: The Apple Mafia[6][7][8] as well as various phreaking/hacking groups, the most high-profile being the Knights of Shadow. When KOS fell apart after a series of arrests, many of the surviving members were absorbed into Kroupa's final group affiliation: the Legion of Doom (LoD/H).[9]

Kroupa started publishing some of his hacking techniques when he would have been around 12 or 13.[10] There is a significant progression through years of text, which captures Kroupa's early evolution and skills,[11] culminating in an extensive, programmable phone phreaking and hacking toolkit for the Apple II computer, called Phantom Access (which is where the name Phantom Access Technologies, the parent corporation behind MindVox, would later come from).

10.2 The MindVox Years

10.2.1 Voices in my Head (1991–1996)

In the late 1980s and early 1990s, the computer underground had suffered through a series of protracted raids by the Secret Service and FBI, called Operation Sundevil and Operation Redux. Many Legion of Doom members were raided and charged.[12][13][14][15][16] This happened against the backdrop of the first and largest gang war that ever took place in cyberspace, the Great Hacker War between LOD and their rival gang MOD (Masters of Deception).

Considering Kroupa and Fancher's backgrounds and the fact that MindVox employed a motley collection of convicted felons like security expert Len Rose[17] and the infamous Phiber Optik (Mark Abene) who was awaiting a Manhattan grand jury indictment, these were very real issues at the time.

This is the environment in which Patrick Kroupa and Bruce Fancher, launched MindVox. In the words of Bruce Fancher:

> Our greatest fear wasn't whether or not we'd be successful as a company, that was secondary. What concerned us was that one day the Secret Service would kick in the door and just confiscate everything.

This is also the time during which Patrick Kroupa wrote, **Voices in my Head**, *MindVox: The Overture*. Kroupa wrote about the cultural forces that were at play in the hacker underground during the decade that pre-dated the launch of MindVox, considered by some the "Golden Age" of cyberspace.

In the process of writing and releasing *Voices*, Patrick Kroupa stepped out from behind Lord Digital. Instead of status in the hacker underground and notoriety in a sub-culture, Kroupa was being written about as the Jim Morrison of cyberspace[18] and receiving accolades from the mainstream press.[19][20][21][22]

Voices helped define what MindVox became, a counter-cultural media darling meriting full-length features in magazines and newspapers such as Rolling Stone, Forbes, The Wall Street Journal, The New York Times and The New Yorker. *Voices in my Head* was the spark that propelled Kroupa out of obscurity and into the mainstream.

There is no single article that captures this as well as *Sassy* magazine's effusive coverage of MindVox. The long strange trip that began in the hardcore hacker underground, had landed in the middle of a glossy mainstream magazine targeted at an audience of teenage girls, with Kroupa and Fancher displacing that issue's "Cute boy band alert!" with the "Cute cyberpunk alert!".[23]

10.2.2 MIA / DOA (1996–2000)

A running theme through nearly all of Kroupa's writing is his drug use. He was a very vocal proponent of self-selecting one's own state of consciousness and freely wrote and talked about his own drug history. The caveat being, *some* of his drug use was open and public. The fact that he was an advocate of LSD and other psychedelic drugs was no big secret. The darker side of his life — that he regularly lost weeks of time injecting speedballs, was in and out of detoxes and rehabs, and suffered from bipolar disorder — were not publicized or mentioned until nearly a decade later.

Kroupa wrote with great honesty and passion about a variety of topics, but he very carefully danced around his own increasing dependence on heroin. Everybody knew that Kroupa occasionally used heroin, cocaine and dozens of other drugs, but not the extent.

By 1996, MindVox was at the absolute height of its powers, yet it was disintegrating. Bruce Fancher was suddenly part of two or three other start-ups, system repairs that should have taken hours, dragged on for weeks. While the user-base kept growing, the previously high level of intelligent discourse within the internal conferences had suffered, and while MindVox was getting more press than ever, all of it read like the same story being retold for the umpteenth time.

Sometime in early-to-mid 1996, Kroupa simply vanished. Freedom of choice, gave way to the downward spiral of hardcore heroin addiction and dysfunction. In his 2005 book, *Hip: The History*, *New York Times* reporter and former *Details* editor John Leland would write:

> In truth far too many of the celebrated figures in these pages led melancholy and difficult lives of isolation, mental illness and drug addiction. Interesting and romantic to read about, but very tough on those who live them.

Kroupa's exact whereabouts and activities from early 1996 until December 1999, remain unknown. He has acknowledged that he travelled throughout North America and spent time living in Mexico, Belize, Puerto Rico, the Czech Republic and eventually Bangkok, Thailand.

The dot.com success story that began with MindVox, eventually hit rock bottom when Patrick Kroupa turned thirty incarcerated, "doin' Cold turkey on cement, in The Tombs".[24] Several months after this arrest, Kroupa finally kicked heroin through the use of the hallucinogenic drug, ibogaine. He was detoxed for the last time in the West Indies, on the Caribbean island of St. Kitts by Dr. Deborah Mash in late 1999.

He subsequently spent four months living at the controversial Buddhist temple, Wat Tham Krabok (which has since been shut down by the Thai government and wrapped in concertina wire, on suspicion of being an international heroin smuggling conduit).[25]

10.3 21st century

A heroin-free Kroupa returned to the United States from Thailand in 2000, and became CTO of Dr. Deborah Mash's Ibogaine Research Project[26] at the University of Miami's Leonard M. Miller School of Medicine.

During the next several years Kroupa appeared in a series of ibogaine-related news reports which aired on television, radio and print media.[27] The most famous example probably being San Francisco's KRON news-report, which aired in 2004 and features Kroupa and Mash in a ten-minute long pro-ibogaine story.[28]

Kroupa is regularly a featured speaker at psychedelics and harm reduction conferences.[29][30][31][32][33][34] He seems to have a penchant for appearing at speaking engagements with multiple cups of coffee lined up in front of him, sometimes chain-smoking cigarettes through hour-long presentations. Whatever ibogaine has done for his other addictions, it seems to have had no effect on his nicotine and caffeine dependence.[35][36][37]

10.3.1 Yippies and the counterculture

While Kroupa's past history with the Yippies began at around age 13 or 14,[2] when the Yippies formalized a Yippie Speakers Bureau in 2003, consisting of: Paul Krassner, Dana Beal, Robert Altman, Grace Slick, Stew Albert, Dennis Peron, Ed Rosenthal, Jack Hoffman, Steven Conliff and Hunter S. Thompson, and went on tour during 2003-2004, the line-up featured the surprising inclusion of former Black Panther Party leader Dhoruba bin Wahad, and Patrick Kroupa, who wasn't born when the Yippies first became a cultural force in the United States, and was 2-3 generations younger than his closest compatriot.[38] It is unknown whether the YSB remains active; it went on hiatus in 2005 with the deaths of Stew Albert, Steven Conliff, and suicide of Hunter S. Thompson.

On November 15, 2007, he spoke at the University Philosophical Society (Trinity College, Dublin), discussing ibogaine, the worldwide War on Drugs, and advocating the legalisation of all narcotics.[39] The following Monday (November 19, 2007) Kroupa appeared on Ireland's national television station TV3's Ireland AM talk show, calling the War on Drugs:

> ...an unequivocal, catastrophic, world-wide failure, that has destroyed countless lives, set fire to hundreds of billions of dollars and produced no discernible results. There is no lack of drugs, basically, anywhere on planet Earth. The number of people using drugs has not decreased. While the street price of drugs hasn't gone up, the purity levels have steadily risen. But hey, we sure do have a lot of people in prison!

Kroupa is High Priest in the Eastern European based Sacrament of Transition[40] (a religious organization whose initiation rituals involve the sacramental use of ibogaine), and a member of Cult of the Dead Cow.[41]

10.4 Bibliography

10.4.1 Essays

- **Voices In My Head** *MindVox: The Overture* (1992), Patrick K. Kroupa. , ,

10.4.2 Magazines

- The Akashic Records of Cyberspace (1993), Patrick K. Kroupa. Mondo 2000.
- Memoirs of a Cybernaut (1993), Patrick K. Kroupa. Wired.
- Agr1pPa - A Book of The Mentally Disturbed (1993), Patrick K. Kroupa. Mondo 2000. ,
- The Secret Service is Neither (1994), Patrick K. Kroupa. Mondo 2000.
- Heroin Times: Ibogaine Series (2000–2003), Patrick K. Kroupa. Heroin Times.

10.4.3 Medical journals

- Ibogaine: Treatment Outcomes and Observations (2003), Hattie Wells (Epoptica) & Patrick K. Kroupa (Junk the Magic Dragon), MAPS (Multidisciplinary Association for Psychedelic Studies, Volume XIII, Number 2).
- Ibogaine in the 21st Century: Boosters, Tune-ups and Maintenance (2005), Patrick K. Kroupa & Hattie Wells. MAPS (Multidisciplinary Association for Psychedelic Studies, Volume XV, Number 1).

10.5 References

10.5.1 Books

- Rudy Rucker & R. U. Sirius, (1992) *User's Guide to the New Edge.* (ISBN 0-06-096928-8)
- Bruce Sterling, (1993) *The Hacker Crackdown: Law And Disorder On The Electronic Frontier.* (ISBN 0-553-56370-X)
- Philip Bacweksi, Tod Foley, and Billy Barron (1994) *Tricks of the Internet Gurus.* (ISBN 0-672-30599-2)
- Frank Biocca, Mark R. Levy, (1994) *Communication in the Age of Virtual Reality.* (ISBN 0-8058-1550-3)
- J C Herz, (1995) *Surfing on the Internet.* (ISBN 0-316-36009-0)
- St. Jude (Jude Milhon), (1995) *The Real Cyberpunk Fakebook.* (ISBN 0-679-76230-2)
- Jeff Goodell, (1996) *The Cyberthief and the Samurai.* (ISBN 0-440-22205-2)
- Charles Platt, (1997) *Anarchy Online.* (ISBN 0-06-100990-3)
- Melanie McGrath, (1998) *Hard, Soft & Wet* (ISBN 0-00-654849-0)
- Richard Power, (2000) *Tangled Web: Tales of Digital Crime from the Shadows of Cyberspace.* (ISBN 0-7897-2443-X)

10.5. REFERENCES

- Rebecca Gurley Bace, (2000) *Intrusion Detection*. (ISBN 1-57870-185-6)
- John Biggs, (2004) *Black Hat*. (ISBN 1-59059-379-0)
- Joseph M. Kizza, (2005) *Computer Network Security*. (ISBN 0387204733)
- John Leland, (2005) *Hip: The History*. (ISBN 0-06-052817-6)

10.5.2 Magazines and newspapers

- Forbes, William Flanagan (1992), The Playground Bullies Have Learned to Type
- Mondo 2000, Andrew Hawkins (1992), There's A Party in my Mind... MindVox!
- Associated Press, Frank Bajak (1993), Wiring the Planet: MindVox!
- The New Yorker (1993), CyberHero
- Wired Magazine, Charles Platt (November 1993), MindVox: Urban Attitude Online
- Sassy Magazine, Margie Ingall (1993), Hi Girlz, See You in Cyberspace!
- New York Magazine, Jeff Goodell (1994), Boot Up and See Me Sometime
- NY Times, John Leland (May 1, 2003), Yippies' Answer to Smoke-Filled Rooms
- Ocean Drive, Tristram Korten (2006), A Cure for Addiction?
- Radar, Tristram Korten (October/November 2008), The Electric Acid Kool-aid Cure

10.5.3 Medical journals

- Brian Vastag, **Addiction Treatment Strives for Legitimacy** JAMA (Journal of the American Medical Association Vol. 288 No. 24, December 25, 2002)

10.5.4 Public Access U.S. government documents

- United States. Congress. Senate. Committee on Governmental Affairs. Permanent Subcommittee on Investigations, (1996). Security in Cyberspace: Hearings before the Permanent Subcommittee on Investigations of the Committee on Governmental Affairs, United States Senate, One Hundred Fourth Congress, Second Session, May 22, June 5, 25, and July 16, 1996

    Available from U.S. G.P.O., Supt. of Docs., Congressional Sales Office. (ISBN 0-16-053913-7)

10.5.5 Film

- Benjamin De Loenen (2005) *Ibogaine: Rite of Passage*. LunArt Productions iMDB

10.5.6 Television

- KRON (2004). Hallucinogen May Cure Drug Addiction

10.5.7 Radio

- KNX 1070 News Radio (2005). Ibogaine

10.5.8 Music

- Billy Idol (1993) *Cyberpunk*, EMI

10.6 Notes

[1] Zdeněk Kroupa 1921-1999

[2] Internet Gurus Tod Foley

[3] Blacklisted News: A Secret History of the 80's Yippie Book Collective. Bleecker Publishing (1984)

[4] The First Trinity: the Commodore PET, the Radio-Shack TRS-80, and the Apple (1977-1980)

[5] The Columbia Electronic Encyclopedia, 6th edition

[6] The Apple Mafia Story

[7] Apple Mafia Krack title page 1

[8] Apple Mafia Krack title page 2

[9] THE HACKER CRACKDOWN: Law and Disorder on the Electronic Frontier War on The Legion, Bruce Sterling

[10] A Guide To ADS Systems, Lord Digital (1982)

[11] RSX11M Version 3.X Real Time Operating System Terminus and Lord Digital (1984)

[12] THE HACKER CRACKDOWN: Law and Disorder on the Electronic Frontier Sting Boards, Bruce Sterling

[13] CS/EP142 Computers and Society, 1996

[14] Computer Underground Digest Volume 2, Issue #2.16 (December 10, 1990)

[15] Operation Sun-Devil Phrack Magazine, Issue: 32, Article: 10

[16] International Intrusions: Motives and Patterns Kent E. Anderson

[17] Boardwatch Magazine: MindVox 1992

[18] Surfing on the Internet J. C. Herz (ISBN 0-316-36009-0)

[19] MindVox: Urban Attitude Online. Wired Magazine, 1993, Charles Platt

[20] Wiring the Planet-MindVox! Frank Bajak, Associated Press, 1993

[21] Boot Up and See Me Sometime New York Magazine, 1994

[22] There's A Party in my Mind... MindVox! Andrew Hawkins, Mondo 2000, 1993

[23] Hi Girlz! See you in Cyberspace Sassy Magazine, 1994.

[24] Sound Bites of Patrick Kroupa, at the Drug Policy Alliance conference DPA, New Orleans, 30 December 2007

[25] Yearning to be Hmong

[26] Chief Technology Officer, Ibogaine Research Project

[27] MindVox's Ibogaine media section

[28] Hallucinogen May Cure Drug Addiction KRON, 2004

[29] Psychedelic Television, 2006 Ibogaine Conference

[30] International Ibogaine conference listings

[31] American Association for the Treatment of Opioid Dependence - 5th National Harm Reduction Conference

[32] Drug Policy Alliance, 2003. Life in the Psychedelic Ghetto, Patrick Kroupa

[33] New York City Ibogaine Forum 2005 - Ibogaine Low Dose and Maintenance Therapy

[34] NYC Ibogaine and Iboga Forum, 2003

[35] Daniel Pinchbeck, Sandra Karpetas, Patrick Kroupa with coffee-cups, Ibogaine conference, 2003

[36] Sandra Karpetas, Patrick Kroupa with coffee-cups, Ibogaine conference, 2003

[37] Kroupa chain-smoking cigarettes, with multiple coffee-cups, Kiblix 2003, Linux security conference in European Union

[38] Yippie Speaker's Bureau

[39] The Irish Examiner, Legalisation of narcotics up for debate November 15, 2007

[40] Sacrament of Transition

[41] Cult of the Dead Cow, Introducing two new members! Feb 19, 2006

10.7 External links

- Personal homepage
- Phantom Access Exhibit
- "Wonderful Things" – War On Drugs essay
- Textfiles List of Losers, 1984
- Patrick K. Kroupa at the Internet Movie Database

Chapter 11

Elias Ladopoulos

Elias Ladopoulos[1] is a technologist and investor from New York City.[2] Under the pseudonym **Acid Phreak**,[3] he was a founder of the *Masters of Deception* (MOD) hacker group[4] along with Phiber Optik (Mark Abene) and Scorpion (Paul Stira). Referred to as The Gang That Ruled Cyberspace[5] in a 1995 non-fiction book, MOD was at the forefront of exploiting telephone systems to hack into the private networks of major corporations.[6] In his later career, Ladopoulos developed new techniques for electronic trading and computerized projections of stocks and shares performance, as well as working as a security consultant for the defense department. He is currently CEO of Supermassive Corp, which is a hacker-based incubation studio for technology start-ups].[7]

11.1 Founding of MOD

When Ladopoulos and Stira were engaged in exploring an unusual telephone system computer, Ladopoulos suggested seeking advice from Phiber Optik (Mark Abene), a well-known phreak who was also a member of the prestigious Legion of Doom (LOD) group. A productive phone hacking partnership developed, with the group later branding themselves Masters of Deception (MOD).[8]

MOD's hacking exploits included taking control of every major phone system and global packet-switching network in the United States. Ladopoulos claims that he and another hacker were able to place a call to Queen Elizabeth II of England. Their pranks included taking over the printers of the Public Broadcasting Service (PBS), an incident that escalated when another hacker used the access they had established to wipe the PBS systems. The group is also known for retrieving phone and credit information for celebrities such as Julia Roberts and John Gotti.[9]

11.1.1 Conflict with former Legion of Doom members

Abene's involvement in both LOD and MOD showed a natural alignment between the two groups in MOD's early years. As LOD's original membership broke up however, conflicts arose between Abene and Eric Bloodaxe (Chris Goggans), another LOD member. Goggans declaring that Abene had been expelled from LOD, resulted in a permanent split between the two groups. Ladopoulos is credited with writing "The History of MOD" for "other hackers to envy."[10] Further disagreements and pranks, including the hacking of Goggans's security consultancy ComSec,[11] have been characterized as the Great Hacker War.[12]

11.1.2 Prosecution

On January 15, 1990 (Martin Luther King Day), the AT&T telephone network crashed.[13] Later investigations revealed the cause to be a software bug, however an FBI task force that had been investigating MOD was convinced the group was implicated. On January 24 the FBI raided the homes of five MOD members, including Ladopoulos, Abene and Stira.[14]

Despite being released without charge due to lack of evidence, the MOD members were later re-arrested on a conspiracy charge following wire-tapping of future MOD members. After Abene rejected a plea bargain, Ladopoulos refused to testify against his fellow hacker, pleaded guilty and was sentenced to 6 months in a supervised camp facility, followed by 6 months house arrest. According to U.S. Attorney Otto Obermaier it was the "first investigative use of court-authorized wiretaps to obtain conversations and data transmissions of computer hackers" in the United States.

11.2 Career

After completing his sentence, Ladopoulos was hired as a security engineer by the Reuters-owned electronic trading business, Instinet. Hiring other former hackers, Ladopoulos built a department responsible for securing Instinet's global trading operations and developing security systems that were later acquired by NASDAQ. Later, as a consultant for Instinet, Ladopoulos also worked as VP Operations for the government security contractor NetSec (later Verizon Government).

In 2008, he founded Kinetic Global Markets with Roger Ehrenberg. As CEO and CIO, he led a team pioneering new approaches to systematic trading based on the computational analysis of terms used in SEC filings. Ladopoulos consulted on Ehrenberg's launch of IA Venture Capital.

In 2013, Ladopoulos founded Supermassive Corp.,[15] which describes itself as the original hacker incubation studio, "bringing together teams of extremely unique talents to rapidly prototype ideas that have a big impact."

11.3 References

[1] Guisnel, Jean (1997). *CyberwarsL Espionage on the Internet*. Basic Books. p. 118. ISBN 0-7382-0260-6.

[2] "Raptor Bites Into VC Market With $32M Toward Fund I | Raptor Group". *www.raptorgroup.com*. Retrieved 2016-04-11.

[3] McMullen, John. "Reflections On Hacker Sentencing 07/29/93". *https://w2.eff.org*. External link in |website= (help)

[4] Ramirez, Mary B. W. Tabor With Anthony (1992-07-23). "Computer Savvy, With an Attitude; Young Working-Class Hackers Accused of High-Tech Crime". *The New York Times*. ISSN 0362-4331. Retrieved 2015-12-29.

[5] Slatalla, Michelle; Quittner, Joshua (10 January 1996). *Masters of Deception: The Gang That Ruled Cyberspace*. Harper Collins. ISBN 978-0-06-092694-6.

[6] Auza, Jun. "7 Most Notorious Computer Hacker Groups of All Time".

[7] http://www.usatoday.com/story/tech/personal/2015/02/04/ozy-hacker-proof-helpers/22829861/

[8] "Gang War in Cyberspace". *WIRED*. Retrieved 2015-12-28.

[9] "Hacker Sentenced for Computer Crimes". November 5, 1993. Retrieved Dec 29, 2015.

[10] "NYTimes". *www.nytimes.com*. Retrieved 2015-12-29.

[11] "Computer Savvy, With an Attitude: Young Working-Class Hackers Accused of High-Tech Crime". *partners.nytimes.com*. Retrieved 2015-12-29.

[12] "Gang War in Cyberspace". *WIRED*. https://plus.google.com/+WIRED. Retrieved 2015-12-28. External link in |publisher= (help)

[13] Desai, Manthan. *Hacking for Beginners: a beginners guide to learn ethical hacking*. hackingtech.co.tv. pp. 272, 274.

[14] Sterling, Bruce (1992). *The Hacker Crackdown*. New York: Bantam. p. 233. ISBN 9780553080582.

[15] http://www.usatoday.com/story/tech/personal/2015/02/04/ozy-hacker-proof-helpers/22829861/

-

11.4 External links

- The History of MOD
    - modbook1.txt — *"The History of MOD: Book One: The Originals"*
    - modbook2.txt — *"The History of MOD: Book Two: Creative Mindz"*
    - modbook3.txt — *"The Book of MOD: Part Three: A Kick in the Groin"*
    - modbook4.txt — *"The Book of MOD: Part Four: End of '90-'1991"*
    - modbook5.txt — *"The Book of MOD: Part 5: Who are They And Where Did They Come From? (Summer 1991)"*

- Small Scale Sin, Act Three http://www.thisamericanlife.org/radio-archives/episode/2/small-scale-sin?act=3#play

Chapter 12

Lucky225

Lucky225, a.k.a. **Jered Morgan**, is a Southern California phone phreak and white hat security professional. He is most known for his social engineering abilities, co-hosting internet radio show Default Radio and exploration and knowledge of caller ID spoofing, Calling Party Number (CPN), and Automatic Number Identification (ANI).

Lucky225 has been published several times in 2600 Magazine, K-1ine Magazine, Von Magazine, and other circulars. He's been mentioned in Wired, Hack Canada, and on CNN. As a privacy advocate he has written and spoken about how to do almost anything without using your Social Security Number. He's also spoken at several conferences, including Defcon and H.O.P.E.

12.1 See also

- Mojave phone booth

12.2 Notable Links

- Default Radio
- Wired, an independent media source, has this article mentioning Lucky and the Default Radio boys, interviewed after a speech at Defcon
- Kevin Poulsen found the information Lucky225 had in 2004 new at the time
- rootsecure independently reports about Lucky225's K7 voicemail found in Paris Hilton's sidekick notes.
- Doug TV - An Internet TV show he was the star from 2004 til the end of the show in 2005. His last appearance for the returning of Doug TV was filmed in July 2007 which is said to be his last official time on the series.
- - How Good Is the Blind Hacker? Mitnick Weighs In, and FBI's Hunt for Lil' Hacker.

Chapter 13

Phone hacking

This article is about the use of telephone technology to steal information. For the manipulation of telephone call routing, see Phreaking.

Phone hacking is the practice of intercepting telephone calls or voicemail messages, often by accessing the voicemail messages of a mobile phone without the consent of the phone's owner. The term came to prominence during the News International phone hacking scandal, in which it was alleged (and in some cases proved in court) that the British tabloid newspaper the *News of the World* had been involved in the interception of voicemail messages of the British Royal Family, other public figures, and the murdered schoolgirl Milly Dowler.[1]

13.1 Risks

Although many mobile phone user may be targeted, "for those who are famous, rich or powerful or whose prize is important enough (for whatever reason) to devote time and resources to make a concerted attack, it is usually more common, there are real risks to face."[2]

13.2 Techniques

13.2.1 Voicemail

Contrary to what to their name suggests, scandals such as the News International phone hacking scandal have little to do with hacking phones, but rather involve unauthorised remote access to voicemail systems. This is largely possible through weaknesses in the implementations of these systems by telcos.[3]

A simple and effective hack against a pabx system is to attempt to call a direct dial number with voicemail and attempt to enter the voicemail features by entering a weak password while the voicemail initial greeting is being played. If the hacker manages to guess the right password, the pabx may have a "call me back" function. The hacker then selects the call back function, but enters a premium rate number for the call back. The pabx calls back the premium rate line, monetising the attack for the hacker. It is important to turn off the call back feature on the pabx, to use strong passwords. Note that automation techniques will then exploit the hack by constantly calling the compromised voicemail and entering the premium rate number ad-Infinitum.

Since the early days of mobile phone technology, service providers have allowed access to the associated voicemail messages via a landline telephone, requiring the entry of a Personal Identification Number (PIN) to listen to the messages. Many mobile phone companies used a system that set a well-known four digit default PIN that was rarely changed by the phone's owner, making it easy for an adversary who knew both the phone number and the service provider to access the voicemail messages associated with that service.[4] Even where the default PIN was not known, social engineering could

13.2. TECHNIQUES

Phone hacking often involves unauthorized access to the voicemail of a mobile phone.

be used to reset the voicemail PIN code to the default, by impersonating the owner of the phone during a call to a call centre.[5][6] Many people also use weak PINs that are easily guessable; to prevent subscribers from choosing PINs with weak password strength, some mobile phone companies now disallow the use of consecutive or repeat digits in voicemail

PIN codes.[7]

During the mid-2000s, it was discovered that calls emanating from the handset registered against a voicemail account were put straight through to voicemail without the caller being challenged to enter a PIN. An attacker could therefore use caller ID spoofing to impersonate a victim's handset phone number and thereby gain unauthorized access to the associated voicemail without a PIN.[8][9]

Following controversies over phone hacking and criticism that was levelled at mobile service providers who allowed access to voicemail without a PIN, many mobile phone companies have strengthened the default security of their systems so that remote access to voicemail messages and other phone settings can no longer be achieved via a default PIN.[4] For example, AT&T announced in August 2011 that all new wireless subscribers would be required to enter a PIN when checking their voicemail, even when checking it from their own phones, while T-Mobile stated that it "recommends that you turn on your voice mail password for added security, but as always, the choice is yours."[10]

13.2.2 Handsets

An analysis of user-selected PIN codes suggested that ten numbers represent 15% of all iPhone passcodes, with "1234" and "0000" being the most common, with years of birth and graduation also being common choices.[11] Even if a four-digit PIN is randomly selected, the key space is very small (10^4 or 10,000 possibilities), making PINs significantly easier to brute force than most passwords; someone with physical access to a handset secured with a PIN can therefore feasibly determine the PIN in a short time.[12]

Mobile phone microphones can be activated remotely by security agencies or telcos, without any need for physical access.[13][14][15][16][17][18] This "roving bug" feature has been used by law enforcement agencies and intelligence services to listen in on nearby conversations.[19]

Other techniques for phone hacking include tricking a mobile phone user into downloading malware which monitors activity on the phone, or bluesnarfing, which is unauthorized access to a phone via Bluetooth.[6][20]

13.2.3 Other

There are flaws in the implementation of the GSM encryption algorithm that allow passive interception.[21] The equipment needed is available to government agencies or can be built from freely available parts.[22]

In December 2011, German researcher Karsten Nohl revealed that it was possible to hack into mobile phone voice and text messages on many networks with free decryption software available on the Internet. He blamed the mobile phone companies for relying on outdated encryption techniques in the 2G system, and said that the problem could be fixed very easily.[23]

13.3 Legality

Phone hacking, being a form of surveillance, is illegal in many countries unless it is carried out as lawful interception by a government agency. In the News International phone hacking scandal, private investigator Glenn Mulcaire was found to have violated the Regulation of Investigatory Powers Act 2000. He was sentenced to six months in prison in January 2007.[24] Renewed controversy over the phone hacking claims led to the closure of the *News of the World* in July 2011.[25]

In December 2010, the Truth in Caller ID Act was signed into United States law, making it illegal "to cause any caller identification service to knowingly transmit misleading or inaccurate caller identification information with the intent to defraud, cause harm, or wrongfully obtain anything of value."[26]

13.4 See also

- Mobile security

- Telephone tapping
- Operation Weeting
- Phreaking

13.5 References

[1] Davies, Nick; Hill, Amelia (4 July 2011). "Missing Milly Dowler's voicemail was hacked by News of the World". *The Guardian*. Retrieved 13 July 2011.

[2] Wolfe, Henry B (February 2010). "Mobile Phone Security" (PDF). *The TCSM Journal*. **1** (2): 3.

[3] Rogers, David (7 July 2011). "Voicemail Hacking and the 'Phone Hacking' Scandal - How it Worked, Questions to be Asked and Improvements to be Made". Copper Horse Solutions. Retrieved 25 Jul 2012.

[4] "Who, What, Why: Can Phone Hackers Still Access Messages?". BBC News. 6 July 2011.

[5] Voicemail hacking: How Easy Is It?, *New Scientist*, 6 July 2011

[6] Milian, Mark (8 July 2011). "Phone Hacking Can Extend Beyond Voice Mail". *CNN*. Retrieved 9 July 2011.

[7] Grubb, Ben (8 July 2011). "Vulnerable voicemail: telco-issued PINs insecure". *The Sydney Morning Herald*. Retrieved 9 July 2011.

[8] Cell Phone Voicemail Easily Hhacked, *MSNBC*, 28 February 2005

[9] Kevin Mitnick Shows How Easy It Is to Hack a Phone, interview with Kevin Mitnick, *CNET*, 7 July 2011

[10] Soghoian, Christopher (9 August 2011). "Not an option: time for companies to embrace security by default". *Ars Technica*. Retrieved 25 July 2012.

[11] Rooney, Ben (15 June 2011). "Once Again, 1234 Is Not A Good Password". *The Wall Street Journal*. Retrieved 8 July 2011.

[12] Greenberg, Andy (27 Mar 2012). "Here's How Law Enforcement Cracks Your iPhone's Security Code". *Forbes.com*. Retrieved 25 Jul 2012.

[13] Schneier, Bruce (December 5, 2006). "Remotely Eavesdropping on Cell Phone Microphones". *Schneier On Security*. Retrieved 13 December 2009.

[14] McCullagh, Declan; Anne Broache (December 1, 2006). "FBI taps cell phone mic as eavesdropping tool". *CNet News*. Retrieved 2009-03-14.

[15] Odell, Mark (August 1, 2005). "Use of mobile helped police keep tabs on suspect". *Financial Times*. Retrieved 2009-03-14.

[16] "Telephones". *Western Regional Security Office (NOAA official site)*. 2001. Retrieved 2009-03-22.

[17] "Can You Hear Me Now?". *ABC News: The Blotter*. Retrieved 13 December 2009.

[18] Lewis Page (2007-06-26). "'Cell hack geek stalks pretty blonde shocker'". The Register. Retrieved 2010-05-01.

[19] Brian Wheeler (2004-03-02). "'This goes no further...'". BBC News Online Magazine. Retrieved 2008-06-23.

[20] How easy is it to hack a mobile?, *BBC News*, 7 September 2010

[21] Jansen, Wayne; Scarfone, Karen (October 2008). "Guidelines on Cell Phone and PDA Security" (pdf). National Institute of Standards and Technology. Retrieved 25 Jul 2012.

[22] McMillan, Robert. "Hackers Show It's Easy to Snoop on a GSM Call". *IDG News Service*.

[23] O'Brien, Kevin J. (25 December 2011). "Lax Security Exposes Voice Mail to Hacking, Study Says". *The New York Times*. Retrieved 28 December 2011.

[24] "Pair jailed over royal phone taps ", *BBC News*, 26 January 2007

[25] News of the World to close amid hacking scandal, *BBC News*, 7 July 2011

[26] Truth in Caller ID Act of 2010, December 22, 2010, accessed 7 July 2011

13.6 External links

- How Phone Hacking Worked and How to Make Sure You're Not a Victim at Sophos
- "Phone hacking collected news and commentary". *The Guardian*.
- Timeline: News of the World phone-hacking row, *BBC News*, 5 July 2011
- Full Q&A On The Phone Hacking Scandal, *Sky News*, 5 July 2011
- Anatomy of the Phone-Hacking Scandal, *The New York Times*, 1 September 2010
- The Rise of Caller ID Spoofing, *The Wall Street Journal*, 5 February 2010
- Phone hacking: Are you safe?, Rory Cellan-Jones, *BBC News*, 12 July 2011

Chapter 14

Phantom Access

Phantom Access 5.7K title page, featuring a confused TIE fighter in front of a Death Star and the post-divestiture AT&T "death star" logo.

Phantom Access was the name given to a series of hacking programs written by Patrick Kroupa (a.k.a., Lord Digital) of LOD. The programs were worked on during the early to mid 80s (1982–1986), and designed to run on the Apple II computer and Apple-Cat modem.

14.1 History

There are a variety of references to the Phantom Access programs in texts from underground Bulletin Board Systems from the 80s,.[1][2][3] Reading the files and messages, it appears that the Phantom Access name was a given to an entire series of programs coded by Lord Digital and apparently for internal LOD use, because the programs were not distributed to the "public" or even most other members of the hacker underground of the time.

Much like Festering Hate, there are references to the programs in a variety of mainstream press articles of the early and mid 90s when MindVox first came online,.[4][5][6] Phantom Access is also where the parent company that launched MindVox, Phantom Access Technologies, took its name from.

14.2 Hacking as art

The overall package is extremely polished and surprisingly professional looking, taking into consideration that Phantom Access was made for internal release, and definitely fell into the black hat category of tools. It was obviously something that would never be commercially released, or apparently, released at all. Yet the attention to detail is stunning. The Phantom Access disks contain graphics, utility programs, extensive, well written and formatted documentation, and sample "sub-modules" which contain instructions for the Phantom Access program itself.

The programs also contain a high level of aesthetics, melting Matrix-style green text screens that spin and fall apart as the next section of the program opens, a collection of strange easter eggs, random poems, and Pink Floyd lyrics. There are endless details contained in the package, which must have taken a tremendous amount of time to add, yet contain nothing that is essential for the program function. After looking at the Phantom Access package as a whole, you are left with the impression of painstaking attention to detail, for no logical reason. It's a package that would never be seen by more than a handful of people from the era, and certainly never sold.

It is quite likely that Phantom Access would have sunk back into mythology and never seen the light of day were it not for the efforts of digital historian and film-maker, Jason Scott, who featured Phantom Access as the first exhibit on textfiles.com, in January 2006.[7] Nearly 20 years after it was written, Phantom Access had made the transition from lost artifact of the past, to finally being made available to anyone who wanted to see what it was all about.[8]

14.3 Phantom Access 5.7K

The Phantom Access disks that were leaked, contained one full Apple II disk side of software and an additional disk of documentation written about the programs.[9] There is another text archive of messages from the era that were apparently posted when Phantom Access was leaked to The Underground BBS in the late 80s.[10]

The programs come on a disk containing something called "ZDOS". As of 2006, there is not a single reference to any disk operating system or variant, for the Apple computer, with any such name. It is unknown if *ZDOS* was ever a commercial product, or something written specifically for Phantom Access. It is also unknown whether ZDOS is itself some kind of virus. Messages from the era indicate that the Phantom Access leak may contain a virus, and taking into account Festering Hate, it is certainly possible.

Phantom Access itself appears to be a highly-programmable common interface, which follows instructions contained in a variety of files. At the topmost level it seems to be a toolkit for utilizing all the special functions of the Apple-Cat modem, it scans systems, hacks codes, functions as a blue box, and exports the results into a series of files which can be manipulated using all the other programs in the series.

A quote from the Phantom Access Documentation:

> *Phantom Access 5.7K is the hacker itself. It could be described as the final processing unit of the instruction sets, but without the utilities it would be useless to the end user, as that is ALL it is. Sub-Modules must already exist prior to usage. This was a necessary compromise, as there is NO memory left on a 64K system once the Slider's and Rotation system are activated.*

It uses EXEC files as a form of primitive scripting for the Apple II. Reading through the messages of the era,[11] the scripts are doing direct writes to various registers and parts of memory using the POKE command. The programs regularly check memory to see what is running or loaded and generally seem to take over control of the computer.

14.4 Final Evolution

Lacking an Apple II computer and Apple-Cat modem, in addition to their historical value, perhaps the most useful and interesting part of the Phantom Access programs is the extensive documentation Kroupa wrote.[12] In addition to explaining how to program the sub-modules, the documents provide an extensive overview of phreaking information, information about the other programs in the Phantom Access series (which appear to have been other system penetration tools and rootkits, before the term "rootkit" existed), and the eventual goal of the whole series, which seems to have been turning the entire Apple II computer and Apple-Cat modem into a programmable phreaking box, which could be plugged into the computers Kroupa and other LOD members were abandoning the Apple platform and switching over to (NeXT, Sun and SGI hardware).[13]

From the Phantom Access documentation:

> *The eventual goal of Phantom Access was to realize a fully automated system for the Apple-Cat modem. The sound sampling and evaluation system has been almost unchanged from revision 4.0 to 5.7 of the series, everything else has been rewritten several times. The final 6.6 revision is a full implementation of the original design (read: it has very little in common with anything in the 5.7 series) with a final processor that is capable of passing data through the Apple-Cat's serial port to an external machine, thus allowing use of the entire Apple computer system as nothing more than a very sophisticated auxiliary modem.*

Towards the late 80s, it looks like Kroupa and LOD had exactly one use left for the Apple II: to utilize the entire computer as a host for the Apple-Cat modem. This makes a very strong statement about how highly valued Novation's, Apple-Cat modem was amongst phone phreaks.

> *This was my solution to working within the Apple's limits. No other modem comes close to having the Apple-Cat's capabilities, but the Apple itself leaves much to be desired.*

14.5 External links

- Phantom Access Exhibit

14.6 References

[1] http://www.totse.com/en/ego/no_laughing_matter/elitecmt.html

[2] http://www.skepticfiles.org/cowtext/100/lozershu.htm

[3] http://p.ulh.as/phrack/p42/P42-03.html

[4] MindVox: Urban Attitude Online Wired Magazine, Volume 1, Issue 5, November 1993

[5] There's a Party in My Mind: MindVox! Mondo 2000, Issue 8, 1992

[6] Trading Data with Dead & Digital by Charles Platt, The Magazine of Fantasy & Science Fiction, 1994

[7] Introduction of Phantom Access Exhibit, by Jason Scott (Retrieved from Textfiles.com)

[8] Phantom Access Exhibit, by Jason Scott (Retrieved from Textfiles.com)

[9] Phantom Access Disk 2 (Documentation) Image (Retrieved from Textfiles.com)

[10] Phantom Access related messages, from the late 80s (Retrieved from Textfiles.com)

[11] http://www.textfiles.com/exhibits/paccess/PhantomAccess.txt

[12] Phantom Access Documentation (converted to text) (Retrieved from Textfiles.com)

[13] Voices in My Head: MindVox The Overture by Patrick Kroupa, 1992

Chapter 15

Phone Losers of America

Warning: Page using Template:Infobox organization with unknown parameter "fgcolor" (this message is shown only in preview).
Warning: Page using Template:Infobox organization with unknown parameter "bgcolor" (this message is shown only in preview).

The **Phone Losers of America** (**PLA**) is an American phreaking group founded in the 1990s, active on the hacking scene. The PLA e-zine was first written in 1990, and the official web site went up in 1994. It has ranked at the top of Alexa's "Prank Call" category. It now however only regularly hosts a podcast called, "The Snow Plow Show" that mainly does prank calls, and occasional phone mobs.[1]

15.1 History

The Phone Losers of America has a history spanning more than 20 years. While the PLA didn't actually exist until 1994, the ideas and articles which were brought to the PLA began in 1992. By 1995, PLA became very well known in the computer underground for their odd contributions to the text zine scene. Most hacker and phreak related bulletin board systems carried the series of PLA text files.

The PLA text files continued until mid-1997, when RBCP suddenly put an end to them. Though the text files ended, the PLA never slowed down. Their website continued to flourish, PLA TV was born, and several prank call CDs were released. The PLA community continued to grow on IRC and on the various PLA Forums. In 2002, RBCP began an internet radio station which he called PLA Radio which still exists today as Cactiradio, along with an IRC channel.(PLA History, 2008)

15.1.1 The e-zine

The PLA e-zine was originally distributed electronically via a dial-up BBS, with an option to buy the magazine and have it shipped by mail. It is not recommended that anyone follow any instructions in any article, since for the most part, the humorous antics (such as Phone fraud) are illegal.

The editor of the e-zine was Brad Carter, who also went by the aliases of RBCP, RedBoxChiliPepper and Alex, the latter being his middle name. Most of the material in the e-zine was written by Brad, especially in the early issues where he was the only person contributing. Eventually he began accepting submissions from readers and the e-zine grew into a multi-article format, much like Phrack and other e-zines of that era. The e-zine also included several ads for breast augmentation devices, fool's gold detectors, and impossible Chinese finger traps.

Each issue of the e-zine usually centered around a single topic, such as BBS hacking, using a red box, revenge tactics, call forwarding hacking, etc. While most of the issue were informative and legitimately taught readers how to do these things, the issues were also heavy on humor and sprinkled with complete nonsense and things that would never work. Much of the fun in reading the PLA e-zine was spotting the nonsense and knowing that many people would take the ridiculous advice seriously. An example of this would be issue #33 (Issue #33, 1995), where PLA provided a phony secret code for pay phones that would cause all of the money to fall out of the change slot.

15.1.2 PLA Radio

In early 2006, RBCP started releasing almost-monthly audio comedy shows called PLA Radio. The shows features comedy skits, commercials, parodies and lots of pranks. The show is released as a podcast and has been featured several times on iTunes and is regularly listed in the Top 10 on Podcast Alley. Some of the names used, like Chris Tomkinson and cactus, are inside jokes for the PLA's dedicated followers. Episodes are released sporadically, sometimes occurring monthly and sometimes only happening a few times a year. They consist of posting fake classified ads and mocking anyone who calls to respond. (PLA Radio, 2008)

15.1.3 PLA TV

Since canceling the PLA zine, RBCP has been regularly releasing short videos, usually revolving around various pranks that him and friends have pulled. Most of the earlier episodes feature RBCP and EvilCal, which includes the legendary Adventures of Elephant and Bird. RBCP still releases videos today, the most recent episodes revolving around modifying a toaster to broadcast on drive-thru frequencies and getting back at the religious people who constantly show up at his door. You can watch most of the PLA TV episodes on YouTube, which are linked from the PLA TV website.

15.1.4 PLA Voice Bridge

The PLA hosts a voice bridge. This bridge was previously hosted on the same extension but the provider changed phone numbers. There are daily conferences from 8pm to 12am EST. The voice bridge is very active during this time, with many people talking and background noise such as soundboards (which many users frown on). The bridge is also a popular place for Telephone Relay Service operators to gather.

A very small number of international enthusiasts have expressed support for an alternative meeting time to allow them to join the fun. (Currently the 8pm to 12am EST schedule occurs in the middle of the night for Europe, the Middle East, and Asia). There has been no official or unofficial secondary meeting time chosen as of March 2007. Most international participants make do with listening in for a short time very late at night or very early in the morning.

Occasionally, a user will create a "secret 800 number" which, when called, forwards to the voice bridge. This is usually accomplished by some method of theft of the toll-free 800 service, and is never officially sanctioned by PLA.[2]

Brad Carter of the Phone Losers currently maintains a telephone network interface for the PLA. This number presents several options, including attempting to hack some answering machines that Brad has, calling the PLA conference line, calling PLA member TheN's home phone, or playing a choose-your-own-adventure story over the phone line.

15.1.5 PLA Community Forums

The Phone Losers web site has always had some kind of forums on their website. In 1997 they used WebForums which were continually attacked by backers of the rival Boulder News Forums. In 1998 and 1999, RBCP used his own forum software in Perl, called FruitWare. These were closed in February 1999 due to abuse.

The site remained without forums for the next couple of years, but communicated with each other on the PLA Email List. In 2002, EvilCal took over by creating Cal's Forums which lasted until he shut them down in April 2006, setting up Cal's Content Kingdom in its place.

Today, the PLA community forums are administrated by RBCP, and were established on 03-21-2006. There are currently over 1,600 registered users, with over 50,000 posts on topics mostly related to the PLA. PLA Forums

Additionally, there is a dedicated PLA Subreddit with active users PLA Subreddit

15.1.6 PLA: The Books

In 2010, the Phone Losers of America book was released. PLA Book The book contains many of the best stories that were found in the original PLA e-zine, such as the Fred Meyer remote intercom pranks, the story of Dino's cordless phone, and the call forwarding experiments. Also included are more recent PLA events, such as RBCP's eBay feedback pranks and the changing of a McDonald's sign. Everything was re-written for the book and some stories are brand new, but based on content originally found in the PLA e-zine.

In 2011, a second Phone Losers of America book was released, called Phone Losers of America: The Complete 'Zine Collection. PLA: The Complete 'Zine Collection This book was simply an archive of all of the original issues of the PLA e-zine, in a mostly unedited format. This book was released "at cost" as a paperback book and for free as an e-book.

15.1.7 Tenth Anniversary

The PLA celebrated their 10th anniversary in 2004. While some of the textfiles show dates as early as 1991, RBCP claims the PLA as its current incarnation didn't fully exist until 1994. Several members of Cal's Forums put their heads together to find a way to celebrate the anniversary in a unique way. Rob Vincent aka Rob T Firefly, Judas Iscariot, Big-E, Liife, I-BaLL, and Murd0c headed up a panel at The Fifth H.O.P.E. conference.

The PLA giving their panel at the 2004 Hackers on Planet Earth convention.

The panel occurred July 9, 2004 at 11pm and was seen by over 800 people in the audience and countless more via streaming media from 2600. Widely regarded as a success, the group played prank videos, held impromptu question and answer sessions, played recordings of prank calls, and nearly caused a riot when giving away free PLA Media CDs. All of the attention caused mentions in USA Today, Wired and The New York Post.

There were also problems when RBCP suggested that Cal (the webmaster of the PLA forums) was planning a hostile takeover of the PLA. Many agreed with this RBCP theory, and many expressed their support when he led a crusade against Cal. But some said RBCP had gone "off his rocker", and was making unfounded threats. The Phone Losers were divided until April 2006, when it was announced that they would lay their differences aside and concentrate instead on pranking people and trying to find "the original spark that united the PLA". (PLA Website Update, 2004)

A second PLA panel was scheduled for July 19, 2008 at 11pm at The Last H.O.P.E.. The presentation included a full multimedia work up of the new pranks the PLA has pulled in the four years since the previous panel, question and answers and live pranks at the hotel. Returning speakers included Murd0c, Rob T. Firefly, I-BaLL and Sidepocket.

15.2 Cactus

A "Cactus" has become the PLA's mascot, as well as catchphrase. The origin of the word dates back to an old prank call by RBCP, where he would say nothing but the word "cactus," over and over. In common usage, It can be stated with a question mark "Cactus?" or as an exclamation "Cactus!" Similar "Cactus" themed prank calls are often made by PLA members. Themed prank calls were often made under the pseudonym, Mildred Monday. (PLA issue #35, 1995)

15.3 Press

The PLA has regularly received a fair amount of attention from the media, beginning with a front page article in the Sunday issue of the Belleville News Democrat on September 3, 1995. (PLA Issue #35, 1995) An editorial was written several days later, followed by another front page article on the PLA a week later. (PLA Issue #38, 1995)

The website and zine received regular writeups in computer magazines such as The Net, The Web and Internet Underground throughout the 1990s. (PLA's press page, 2008) PLA received national attention in 2002, through a segment on Tech TV (YouTube, 2008) and then once again in 2005 when RBCP was interviewed live via satellite on CNBC's On The Money. (YouTube, 2008) PLA was also interviewed in 2005 for an article about Wal-Mart in the Boston Herald. (PLA's press page, 2008)

In August 2015, the Columbia Daily Tribune in Columbia, Missouri featured the PLA in an article about a series of "strange calls" received by local residents who had signed a petition against a crosswalk construction project.[3] The article quoted an FBI representative who allegedly told the *Tribune* that the calls to Columbia residents "would likely be prosecuted on the local level."[3] However, no prosecution of the calls ever occurred.

15.4 United Phone Losers; a PLA Spin-Off

In 1998, Southern California phone phreak "linear" [sic] created a spin-off group of the Phone Losers of America, and dubbed it the United Phone Losers. Though the group reportedly started as a joke, they picked up momentum when RBCP of the PLA recognized the UPL and asked them to take over the now-defunct *PLA Telephone Director0y* (later renamed the *UPL/PLA Telephone Director0y*), a comprehensive list of interesting and/or humorous telephone numbers.

The UPL had a popular message board forum and released 28 issues of its own ezine up until 2002 when they ceased production of the ezine (to return later as sporadic releases). They did continue to host their website, however, and actively maintained their message board and archived ezine and site content until 2004, when the website, http://www.phonelosers.net, was "domain sniped" by cybersquatters. Reclaiming the domain was initially decided against due to the group members' declining interest in continuing the project and focus shifts to other projects and aspects of their lives. However, more recently in 2008, the group and the website have returned (see below).

In 2005, RBCP of the PLA compiled all 28 issues of the defunct UPL ezine and hosted them on the PLA website, stating "So now the tables have turned - the PLA is stealing all of the UPL's material and putting it on their site. This page is a homeage [sic] to the old UPL site and is a complete archive of the UPL issues."

One of UPL's logos, parodying AT&T's old logo

Shortly thereafter, RBCP released an "unauthorized" 29th issue of the United Phone Losers ezine, explaining what happened to the old UPL website as well as including previously unreleased material.

In 2006, linear gained back domain registration of phonelosers.net once the former cybersquatters lost interest and let their registration expire. The site sat still as an archive of the old UPL issues until March 2008 when linear announced that the site and the group had returned and would become active once more.

In June 2008, linear "hacked" PLA Radio and released an "unauthorized" episode with cohost bex0. The episode explained that the take-over was retaliation against RBCP for releasing the issue #29 of the United Phone Losers ezine. In October of the same year, the UPL released the 30th issue of their ezine, bringing it back after about 6 years since a previous "authorized" release.

15.5 The Snow Plow Show

Following the end of The Phone Show and PLA Radio, RBCP (Brad Carter) started hosting a podcast, which he named "The Snow Plow Show". The name is in reference to a joke made in an online classified ad for a snow plow for sale Brad had come across, in which the author questioned the reasoning behind the fact that the words "snow" and "plow" do not rhyme, despite being spelled similarly.

The current host of The Snow Plow Show is solely RBCP, who also records, edits, and releases the podcast via the Phone Losers website, Stitcher, Youtube, and other locations. The style and organization of the show is notable for its more complex, creative, and clever means of carrying out its prank calls.

The calls that take place in the show are all made in a manner that avoids crossing the line of being either illegal or unethical. RBCP avoids making his calls crossing the line of illegality by refusing to make any threats, calls to those aged under 18 (minors in the United States), making a "harassing" number of unwanted phone calls, impersonating law enforcement, causing property damage, inciting physical harm to an individual, or otherwise inciting panic in the recipient which could reasonably cause the person to perform such activities. Besides the above-guidelines summarized that are explicitly followed, RBCP also takes additional care to make the calls ethical by avoiding direct insults, overly-vulgar humor, or allowing any personal details to be publicly revealed. Despite all of these guidelines and principles being carried out, recipients commonly threaten to call the police during the course of the phone call, and sometimes the victim actually makes a call to the police to report the call despite the lack of any illegal activity having had taken place.

As of early 2016, the show often opens with a memorable prank call made on an earlier episode, the official introduction plays, RBCP introduces the show and summarizes the premise of the upcoming show and makes any relevant announcements regarding phones or the show itself. The main part of the show is then played, which may have been either pre-recorded or broadcast to a live audience, many of whom participate in a chat room with RBCP as he makes the calls to suggest ideas and make comments. Lastly, "the voicemails" are played, which are composed of short voicemails left by listeners on a line maintained by RBCP, who responds to each message left. There are several notable characters and listeners that appear on the voicemail fairly regularly, including "Corbin Guy", "Gloria", "Olga", "Crimson", "Brad", "Uncle Scott", and "Ahmed Omadi".

Oftentimes, a business such as an Auto-Repair Shop, Pizza Delivery Restaurant, or 1-800 catalog business is called and made to believe that the prank caller is in fact the respective business' corporate office and that "the orders aren't coming through". They are then told that to resolve the issue, "the IT department" says that the orders can be looked up on their computers by putting in, among other things, the customer's phone number. The employee on the business' phone is then prompted to divulge several of the business' customers' phone numbers, as well as other information about the customer's order. Using this information, prank calls are then made to the customers whose information was just given, the prank-caller often impersonating an employee of the business itself. The customer is often left without any option but to believe that the call is being made from an actual employee of the business because the prank caller often spoofs their phone number to the organization/business' actual phone number as well as reveals information that only the business would know (for example, the person's pizza order, shoe size, or what service was performed on their car).

In another popular type of prank call, listeners of the show from around the United States are encouraged to leave a note on strangers' parked cars stating "Sorry I Dinged Your Car - Roy", along with a provided phone number which, when called, is directed to a voicemail of the fictional "Roy", actually maintained by RBCP. The voicemails that are left are then listened to and the numbers called back by RBCP as "Roy" and commonly told a confusing and/or ridiculous story about how the "ding" was made and/or why the note was left. This style of prank call is now reserved for a single specified period announced once annually for listeners to leave the notes, usually taking place for approximately a month in the fall.

15.6 See also

- MJ Morning Show RBCP's prank calls to MJ et. all are recorded in the PLA media pack.
- Fonejacker
- Tube Bar prank calls
- Pranknet

15.7 References

[1] "Snow Plow Show - Website". *Phone Losers of America*. Retrieved 04-09-15. Check date values in: |access-date= (help)

[2] "PLA's Voice Bridge: 253-397-1819 - Phone Losers of America". *Phone Losers of America*.

[3] "Strange calls about College Avenue project were YouTube prank". *Columbia Daily Tribune*. Retrieved 2016-06-04.

15.8 External links

- Official Phone Losers of America Web Site
- Directory of other PLA State Sites
- Official United Phone Losers Web Site

- Cal's Content Kingdom
- RBCP's Homepage

Chapter 16

Phreaking boxes

Phreaking boxes are devices used by phone phreaks to perform various functions normally reserved for operators and other telephone company employees.

Most phreaking boxes are named after colors, due to folklore surrounding the earliest boxes which suggested that the first ones of each kind were housed in a box or casing of that color. However, very few physical specimens of phreaking boxes are actually the color for which they are named.

Most phreaking boxes are electronic devices which interface directly with a telephone line and manipulate the line or the greater system in some way through either by generating audible tones that invoke switching functions (for example, the blue box), or by manipulating the electrical characteristics of the line to disrupt normal line function (for example, the black box). However a few boxes can use mechanical or acoustic methods - for example, it was possible to use a pair of properly tuned whistles as a red box.

Among the most well-known phreaking boxes were the black box, which tricked switching equipment into believing a call had not been answered when in fact it had, resulting in free incoming long distance calls; the beige box, which is an improvised lineman's handset typically made from a one-piece telephone and alligator clips; the blue box which emulated the in-band signaling tones once used by long distance operators and switching equipment; and the red box, which emulated the tones generated by payphones when coins were deposited.

Today, most phreaking boxes are obsolete due to changes in telephone technology.

16.1 List of phreaking boxes

This is not a comprehensive list. Many text files online describe various "boxes" in a long list of colors, some of which are fictional (parodies or concepts which never worked), minor variants of boxes already listed or aftermarket versions of features (line in use indicators, 'hold' and 'conference' buttons) commonly included in standard multi-line phones.

This list of boxes does not include wiretapping "bugs", pirate broadcasting apparatus or exploits involving computer security.

16.2 See also

- Phreaking

- Dual-tone multi-frequency signaling

16.3 External links

- Glossary of 'box' terms

- The Definitive Guide to Phreak Boxes This is the list that was published on 2600: The Hacker Quarterly, Volume 19 Number 1 (Spring 2002 issue), page 15, on the author's website (ElfQrin.com).

Chapter 17

Plover-NET

Plover-NET, typically mis-denoted as Plovernet, was a popular bulletin board system in the early 1980s.[1][2] Hosted in New York state and originally owned and operated by a teenage hacker who called himself Quasi-Moto,[3] whom was a member of the short lived yet famed Fargo 4A phreak group.[2] The popular bulletin board system attracted a large group of hackers, telephone phreaks, engineers, computer programmers, and other technophiles, at one point reaching over 600 users until LDX, a long distance phone company, began blocking all calls to its number (516-935-2481).[2][4]

17.1 Naming and Creation

The name Plover-NET came from a conversation between Quasi Moto, and Greg Schaefer. The topic of computer games came up. One of them, the 'Extended Adventure' game which was based on the 'Original Adventure' fantasy computer game was mentioned. This game was available on Compuserve and during game play the magic word PLOVER had to be used.

Past sysop of Plover-NET included Eric Corley, under the pseudonym Emmanuel Goldstein, and Lex Luthor, the founder of the notorious hacker group Legion of Doom.[5]

Quasi-Moto personally recounted the creation of Plover-NET,[2]

> I met Lex in person while we lived in Florida during the Fall of 1983 after corresponding via email on local phreak boards. I was due to move to Long Island, New York (516 Area Code) soon after and asked him about starting up a phreak BBS. He agreed to help and flew up during his Christmas break from school in late December 1983. We worked feverishly for a couple of days to learn the GBBS Bulletin Board software which was to run on my Apple with a 300 baud Hayes micoSLOWdom %micromodem% and make modifications as necessary. The system accepted its first phone call from Lex in the first week of January 1984 and it became chronically busy soon after.

17.2 Legion of Doom

Lex Luthor, under the age of 18 at the time, was a COSMOS (Central System for Mainframe Operations) expert, when he operated Plover-NET.[5] At the time there were a few hacking groups in existence, such as Fargo-4A and Knights of Shadow. Lex was admitted into KOS in early 1984, but after making a few suggestions about new members, and having them rejected, Lex decided to put up an invitation only BBS and to start forming a new group. Starting around May 1984 he began using is position on Plover-NET to contact people he had seen on Plover-NET and people he knew personally who possessed the kind of superior knowledge that the group he envisioned should have. He was never considered to be the "mastermind of the Legion of Doom", more the cheerleader and recruiting officer.[5]

17.3 2600

Luthor met 2600 Magazine editor, Emmanuel Goldstein on the Pirates Cove, another 516 pirate/phreak BBS. He invited Goldstein onto Plover and it wasn't long before it became an 'official' 2600 bbs of sorts. When a user logged off the system, a plug for 2600 was displayed with their subscription prices and addresses.[2]

17.4 Operations

The Board initially ran on three apple disk drives with 143 K byte capacity. After a few months of operation, New York hacker Paul Muad'Dib appeared at a TAP meeting being held at "Eddies" in Greenwich Village with a RANA Elite III disk drive in hand. The RANA Elite III had a capacity of about 600 KB which put the total storage capacity of the BBS to just over one Megabyte, fairly large for a phreak board in those days. They gladly accepted the donation but did not ask how he obtained the disk drive. The RANA was later passed on to Lex which he used to house his extensive collection of phreak philes that were available to Legion of Doom BBS users. The location of the overworked RANA is currently unknown although Lex believes he sent it back to Muad'Dib around 1986.

In early 1985 Plover-NET officially closed permanently after Quasi-Moto moved back to Florida and was unable to gain traction in re-establishing the bulletin board when he put it back up after moving.[2]

17.5 References

[1] "The LOD Communications Underground H/P BBS Message Base Project". *Phrack Magazine.* **4** (43): 18 of 27. May 15, 1993. ...to the legendary OSUNY, Plover-NET, Legion of Doom!, Metal Shop, etc. up through the Phoenix Project circa 1989/90. line feed character in |quote= at position 49 (help);

[2] Quasi-Moto (March 1993). "Plover-NET BBS Pro-Phile".

[3] Goldstein, Emmanuel (2010). *Dear Hacker: Letters to the Editor of 2600*. Wiley Publishing, Inc. p. 417. ISBN 978-0-470-62006-9. In the July 1984 issue of 2600, Quasi Moto, sysop of the late Plover-Net BBS..

[4] Dwivedi, Rahul (Feb 2014). *The Core Of Hacking*. RahulDwivediBook.

[5] Gupta, Sandeep (2004). *Hacking in The Computer World*. Mittal Publications. p. 79. ISBN 81-7099-936-7.

Chapter 18

The Secret History of Hacking

The Secret History of Hacking[1] is a 2001 documentary film that focuses on phreaking, computer hacking and social engineering occurring from the 1970s through to the 1990s. Archive footage concerning the subject matter and (computer generated) graphical imagery specifically created for the film are voiced over with narrative audio commentary, intermixed with commentary from people who in one way or another have been closely involved in these matters.

The film starts by reviewing the concept and the early days of phreaking, featuring anecdotes of phreaking experiences (often involving the use of a blue box) recounted by John Draper and Denny Teresi. By way of commentary from Steve Wozniak, the film progresses from phreaking to computer hobbyist hacking (including anecdotal experiences of the Homebrew Computer Club) on to computer security hacking, noting differences between the latter 2 forms of hacking in the process. The featured computer security hacking and social engineering stories and anecdotes predominately concern experiences involving Kevin Mitnick. The film also deals with how society's (and notably law enforcement's) fear of hacking has increased over time due to media attention of hacking (by way of the film *WarGames* as well as journalistic reporting on actual hackers) combined with society's further increase in adoption of and subsequent reliance on computing and communication networks.

John Draper, Steve Wozniak and Kevin Mitnick are prominently featured while the film additionally features comments from or else archive footage concerning Denny Teresi, Joybubbles, Mike Gorman, Ron Rosenbaum, Steven Levy, Paul Loser, Lee Felsenstein, Jim Warren, John Markoff, Jay Foster, FBI Special Agent Ken McGuire, Jonathan Littman, Michael Strickland and others.

18.1 See also

- List of films about computers

18.2 References

[1] The Secret History of Hacking on YouTube

18.3 External links

- *Secret History of Hacking* at the Internet Movie Database
- The Secret History of Hacking on YouTube

Chapter 19

StankDawg

David Blake (born 1971), also known as **StankDawg**, is the founder of the hacking group Digital DawgPound (DDP) and a long-time member of the hacking community. He is known for being a regular presenter at multiple hacking conferences, but is best known as the creator of the "Binary Revolution" initiative, including being the founding host and producer of *Binary Revolution Radio*, a long-running weekly Internet radio show which ran 200 episodes from 2003 to 2007.

19.1 Biography

Blake was born in Newport News, Virginia on September 13, 1971. He received an AAS (Associates in Applied Sciences) degree from the University of Kentucky 1992, and has a BS in Computer Science from Florida Atlantic University as well as a CEH certificate. He presently lives and works as a computer programmer/analyst in Orlando, Florida.[1] Blake is a member of the International High IQ society.[2]

19.2 Hacking

StankDawg is a staff writer for the well-known hacker periodical *2600: The Hacker Quarterly*, as well as the now-defunct *Blacklisted! 411* magazine. He has also been a contributing writer to several independent zines such as *Outbreak*, *Frequency*, and *Radical Future*.[1] He has been a frequent co-host of *Default Radio* and was a regular on *Radio Freek America*, and has appeared on *GAMERadio*, *Infonomicon*, *The MindWar*, *Phreak Phactor*, and *HPR (Hacker Public Radio)*.

He has presented at technology conferences such as DEF CON,[3] H.O.P.E.,[4][5] and Interz0ne.[6][7] He has been very outspoken about many topics, many of which have gotten some negative feedback from different sources. His most controversial article was entitled "Hacking google Adwords" at DefCon13 which drew criticism from such people as Jason Calacanis.[8] among others. His presentation at the fifth H.O.P.E. conference drew some surprise from the AS/400 community.[9]

StankDawg appeared as a subject on the television show The Most Extreme on Animal Planet where he demonstrated the vulnerabilities of wireless internet connections.[10]

Blake chose the handle "StankDawg" in college, where he started a local hacking group which became known as the "Digital DawgPound".[11]

19.3 Digital DawgPound

Main article: Digital DawgPound

The **Digital DawgPound** (more commonly referred to as the "DDP") is a group of hackers, best known for a series of articles in hacker magazines such as *2600: The Hacker Quarterly* and *Make*, the long-running webcast Binary Revolution Radio, and a very active set of forums with posts from high-profile hackers such as Strom Carlson, decoder, Phiber Optik and many more. The stated mission of the DDP is to propagate a more positive image of hackers than the negative mass media stereotype. The group welcomes new members who want to learn about hacking, and attempts to teach them more positive aspects and steer them away from the negative aspects by reinforcing the hacker ethic. Their goal is to show that hackers can, and regularly do, make positive contributions not only to technology, but to society as a whole.[12]

19.3.1 History

The DDP was originally founded and named by StankDawg. His stated reasons were that he had made many friends in the hacking scene and thought that it would be useful to have everyone begin working together in a more organized fashion. He was motivated by the fact that there had been other well-known Hacker Groups in the 1980s who had accomplished great things in the hacking world such as the LoD and the MoD. In 1988, while a junior in high school, StankDawg came up with the name on his way to the "Sweet 16" computer programming competition. He jokingly referred to his teammates as "The Digital Dawgpound".

StankDawg lurked in the shadows of the hacking world for many years throughout college under many different pseudonyms. In 1997 he popped his head out into the public and began becoming more active on IRC and many smaller hacking forums. He saw some people who he thought were insanely brilliant individuals who seemed to have the same mindset and positive attitude towards hacking that he did so he decided to approach a couple of them to see if anyone would be interested in forming a group and working together. There was always a huge emphasis not only on technical competence and variety, but also on strength of character and integrity. The team was a mix of hackers, programmers, phone phreakers, security professionals, and artists. They had experience in multiple programming languages and operating systems. DDP members are not only good programmers and hackers, but more importantly, good people. By 1999 the DDP had its first official members and from this partnership, creativity flowed.

The DDP communicated and worked together on StankDawg's personal site, which was open to anyone who wanted to join in on the fun. StankDawg was never comfortable with the fact that it was his name that was on the domain and that many people who were coming to the site were coming because of his articles or presentations but not really appreciating all of the other great contributions from other community members that were around. In 2002, after watching the web site grow quickly, it was decided that a new community needed to be created for these like-minded hackers who were gathering. This was the start of the biggest DDP project called Binary Revolution which was an attempt at starting a true "community" of hackers. As the site grew, so did the DDP roster.

19.3.2 Members

Over the years, DDP membership has included several staff writers for *2600: The Hacker Quarterly* and *Blacklisted! 411* magazine including StankDawg and bland_inquisitor. They frequently publish articles, provide content, and appear on many media sources across the global Interweb. DDP members are also regular speakers at hacking conferences such as DEF CON, H.O.P.E., Interzone, Notacon, and many more smaller and more regional cons.

Some DDP members hold memberships in Mensa and the International High IQ society.[2] StankDawg is very proud of the diversity of the team and has spoken to this many times on Binary Revolution Radio. Members are from both coasts of the United States to Europe and have even had members from Jamaica, Brazil, and many other countries.

19.3.3 Recognition

The DDP maintains a blog "which they refer to as a "blawg" (Obviously a play on the intentionally misspelled word "Dawg"). Posts by DDP members have been featured on other technology-related sites such as those of Make Magazine,[13][14] HackADay,[15][16] Hacked Gadgets,[17][18] and others.

19.4 Binary Revolution

In 2003, StankDawg moved the forums from his personal site over to a new site as part of a project called the Binary Revolution which he considered a "movement" towards a more positive hacking community.[19]

This "Binary Revolution" is the best known of the DDP projects and is commonly referred to simply as "BinRev". This project was created in an attempt to bring the hacking community back together, working towards a common, positive goal of reclaiming the name of hackers. The Binary Revolution emphasizes positive aspects of hacking and projects that help society. It does this in a variety of outlets including monthly meetings, the weekly radio show Binary Revolution Radio(BRR), a video-based series of shows called HackTV, and very active message board forums.

BinRev is more than just a radio show or forums, although they are certainly the most well-known of the projects. It is actually composed of many parts.

19.4.1 Binary Revolution Radio

Binary Revolution Radio, often shortened to "BRR", was one part of the binrev community. Started and hosted by Blake in 2003, it featured different co-hosts each week, and covered different aspects of hacker culture and computer security.

It was broadcast via internet stream, usually prerecorded in Florida on a weekend, and then edited and released on the following Tuesday, on the DDP Hack Radio stream at 9:30pm EST. Topics included phreaking, identity theft, cryptography, operating systems, programming languages, free and open source software, wi-fi and bluetooth, social engineering, cyberculture, and information about various hacker conventions such as PhreakNIC, ShmooCon, H.O.P.E., and Def Con.

In July 2005 Blake announced that he was going to take a break, and so for the third season, the show was produced by Black Ratchet and Strom Carlson (who had been frequent co-hosts during Blake's run). During the time that they hosted the program, the format rotated between the standard prerecorded format, and a live format which included phone calls from listeners.

Blake returned to the show in May 2006. He maintained the prerecorded format, and brought more community input into the show, by bringing on more members of the Binary Revolution community. For the first episode of the fourth season, BRR had its first ever broadcast in front of live audience during the HOPE 6 convention in New York City, June 2006.[20]

The final episode, #200, took place on October 30, 2007, with a marathon episode which clocked in at 7 hours and 12 minutes.

Notable co-hosts

- Billy Hoffman as ("Acidus")
- Tom Cross (as "Decius")
- Elonka Dunin
- Jason Scott
- Lance James
- Mark Spencer

19.4. BINARY REVOLUTION

- Virgil Griffith
- MC Frontalot
- Lucky225
- Strom Carlson
- Black Rachet

19.4.2 BinRev Meetings

As the forums grew there were many posts where people were looking for others in their area where other hacker meetings did not exist or were not approachable for some reason. Those places that did have meetings were sparse on information. Binary Revolution meetings were started as an answer to these problems and as a place for our forum members to get together. BinRev meetings offer free web hosting for all meetings to help organize the meetings and keep communications alive and to help projects to grow. Some meetings are in large cities like Chicago and Orlando while others are in small towns. Anyone can start their own BinRev meeting by asking in the BinRev forums.

19.4.3 BinRev.net

"BRnet" is the official IRC network of Binary Revolution. It is active at all hours of the day and contains a general #binrev channel but also contains many other channels for more specific and productive discussion.

19.4.4 HackTV

In the middle of 2003, he released an Internet video show entitled "HackTV" which was the first internet television show about hacking, and which has grown into a series of several different shows.[21] They were released irregularly since most of the episodes were filmed by StankDawg in South Florida where he lived at the time. They wanted the show to appear professional in terms of quality, but this made cooperating over the internet difficult. Sharing large video files was difficult and encoded video caused editing problems and quality concerns. The original show was released as full-length 30 minute episodes. This was also a problem since it because more and more difficult to get enough material for full-length episodes. There was also some content that was related to hacking only on a fringe level and StankDawg did not feel it was appropriate to include in the show. This led to other ideas.

HackTV:Underground

In light of the difficulties of putting together the full HackTV original show, and in an attempt to make the show more accessible for community contributions, StankDawg launched a new series that was less focused on format and video quality that focused more on content and ease of participation. This series was titled "HackTV:Underground" or "HTV:U" for short. This series allowed anyone to contribute content in any format and at any length or video quality. The allowed people to film things with basic cameraphone quality video if this was the only way to get the content. One episode of HackTV:U was used by G4techTV show called "Torrent".[22]

HackTV:Pwned

This series of HackTV was a prank style show, similar to the popular "Punk'd" show on MTV at the time. Even the logo is an obvious parody of the Punk'd logo. This series contains pranks that mostly took place at conferences, but is also open to social engineering and other light-hearted content.

19.4.5 DocDroppers

The DocDroppers project is a community project to create a centralized place to store hacking articles and information while still maintaining some formatting and readability. Old ascii text files existed scattered across the internet but they come and go quickly and are difficult to find. They are usually formatted with the very basics and sometimes difficult to read. DocDroppers allows users to submit articles to a centralized place where they can be searchable, easily maintained, and easy to read and reference.

Recently, this project has grown to include encyclopedia style entries on many hacking topics after many were deleted from sites such as Wikipedia. This has caused DocDroppers to include a section on hacker history and culture among its content.

19.5 Selected writing

- "Stupid Webstats Tricks", Autumn 2005, *2600 Magazine*
- "Hacking Google AdWords", Summer 2005, *2600 Magazine*
- "Disposable Email Vulnerabilities", Spring 2005, *2600 Magazine*
- "How to Hack The Lottery", Fall 2004, *2600 Magazine*
- "Robots and Spiders", Winter 2003, *2600 Magazine*
- "A History of 31337sp34k", Fall 2002, *2600 Magazine*
- "Transaction Based Systems", Spring 2002, *2600 Magazine*
- "Batch vs. Interactive", Summer 1999, *2600 Magazine*

19.6 Selected presentations

- "The Art of Electronic Deduction", July 2006, *H.O.P.E. Number Six* (presented again at *Interz0ne 5*, Saturday March 11, 2006)
- "Hacking Google AdWords", July 2005, *DEF CON 13*
- "AS/400: Lifting the veil of obscurity", July 2004, *The fifth H.O.P.E.*

19.7 Projects

Projects that StankDawg was directly involved in creating/maintaining in addition to the ones mentioned above.

- DDP HackRadio - A streaming radio station with a schedule of hacking and tech related shows.
- Binary Revolution Magazine - The printed hacking magazine put out by the DDP.
- Hacker Events - A calendar for all hacking conferences, events, meetings, or other related gatherings.
- Hacker Media - A portal for all hacking, phreaking, and other related media shows.
- Phreak Phactor - The world's first Hacking reality radio show.
- WH4F - "Will Hack For Food" gives secure disposable temporary email accounts.

19.8 References

[1] "Who is StankDawg?". December 11, 2004.

[2] "StankDawg's High IQ Society member page". 2007.

[3] "Defcon 13 speakers list.". August 18, 2005.

[4] "The fifth HOPE speakers". July 2004.

[5] "HOPE number 6 speakers". July 2006.

[6] "Interz0ne 4 speakers". 2004.

[7] "Interz0ne 5 speakers". 2005.

[8] "Hacking Google Adwords - Defcon Panel recap". July 30, 2005.

[9] "Dubious Achievement: iSeries Gets Some Attention From Hackers". July 2004.

[10] "February 15, 2005 episode of The Most Extreme credited as "StankDawg"". February 15, 2005.

[11] "Who is StankDawg?". January 28, 2005.

[12] StankDawg (2004-12-23). "Why Hack?".

[13] Phillip Torrone (2007-04-06). "HOW TO - RFID Enable your front door (with a Parallax BASIC Stamp & 13.5 MHz APSX RW-210)". *Make* Magazine.

[14] Phillip Torrone (2005-06-25). "Apple's Podcasting iTunes 4.9 is out!". *Make Magazine*.

[15] Eliot Phillips (2006-03-27). "Using Radiosondes as cheap GPS trackers". hackaday.com.

[16] Eliot Phillips (2006-07-02). "Email on the Cisco 7960". hackaday.com. Retrieved 2007-07-08.

[17] Alan Parekh (2006-09-18). "RFID Front Door Lock". hackedgadgets.com. Retrieved 2007-07-08.

[18] Alan Parekh (2007-04-03). "RFID Enabling Your Front Door using a Parallax Microcontroller". hackedgadgets.com. Retrieved 2007-07-08.

[19] "Why did you start BinRev?". January 28, 2005.

[20] HOPE Number Six

[21] "Hack TV - Pilot Episode - 08.17.03". August 17, 2003.

[22] Amber MacArthur (May 18, 2006). "Episode 6 - May 18, 2006".

19.9 External links

Other links that were mentioned or referred to in this entry:

- StankDawg's personal site.
- The Digital DawgPound - official site.
- BinRev IRC - Binary Revolution official IRC channel web site & BinRev IRC - Official Binary Revolution IRC network.
- HPR - "Hacker Public Radio" is a daily hacking and technology radio show created by the DDP, infonomicon and others. It has many different hosts.
- BRR Archive - Archive of the hacking radio show presented by members of the DDP (07/2003-10/2007).

- Binary Revolution Meetings - Monthly hacker meetings that encourage participation and offers free hosting for all meetings.
- HackTV - The Internet's first full-length regular Hacking video show.
- Old Skool Phreak - Home of many phreaking related text files.
- RFA Archive - Weekly Radio show about Technology, Privacy and Freedom (02/2002 - 02/2004).

Chapter 20

Dennis Terry

Dennis Dan "Denny" Teresi (born August 14, 1954), now known as **Dennis Terry**, is an American radio disk jockey and former phone phreak most famous for being the person who introduced John Draper to the field of phreaking. Both Draper and Teresi were operating pirate radio stations in the San Jose, California area.[1] Their initial contact came when Teresi responded by telephone to one of Draper's pirate broadcasts.[2]

In the documentary *The Secret History of Hacking*, Teresi is identified as an expert in social engineering.[3] Teresi's mastery of the phone company's jargon allowed him to speak with phone company employees and trick them into revealing more information.

Teresi, who is blind,[4] has hosted an oldies radio show on KSJS since 1976 under the name "Dennis Terry",[5] and for several years he also operated an oldies record store in San Jose.[6]

20.1 References

[1] http://www.afana.org/ktaoroster.htm KTAO roster

[2] http://www.webcrunchers.com/stories/first_visit.html Webcrunchers

[3] http://www.imdb.com/title/tt2338277/fullcredits?ref_=tt_ov_st_sm

[4] "On this day in 1978." *The Spartan Daily*, February 8, 2012, p. 6.

[5] Kava, Brad (December 1, 2006). "Why it sounds a lot like Christmas on the air earlier and earlier". *San Jose Mercury News*. Retrieved December 25, 2014.

[6] http://books.google.com/books?id=ECiBd4mYkVwC&pg=PT315&lpg=PT315&dq=record+store+%22denny+teresi%22&source=bl&ots=U70tr-vRnz&sig=h49QCfNwrc3avtmkae7odqANx5E&hl=en&sa=X&ei=dDqZU6-iDMH5oASktYGYBg&ved=0CDQQ6AEwAw#v=onepage&q=record%20store%20%22denny%20teresi%22&f=false Exploding the Phone: The Untold Story of the Teenagers and Outlaws Who ... By Phil Lapsley

Chapter 21

ToneLoc

Not to be confused with Tone Lōc.

ToneLoc was a popular war dialing computer program for MS-DOS written in the early to mid-1990s by two programmers known by the pseudonyms Minor Threat (Chris Lamprecht) and Mucho Maas. The name ToneLoc was short for "Tone Locator" and was a word play on the name of the rap artist known as Tone Lōc.

ToneLoc took advantage of the extended return codes available on US Robotics modems (e.g., ATX6 [1]) to detect dial tones upon dialing a number and to detect when a human answered the phone in addition to scanning for other modems. Detection of voice numbers sped up the scanning process by disconnecting upon detecting a human instead of timing out waiting for a modem carrier signal. The detection of a dial tone after dialing a number allowed for users to search for poorly secured extenders which could be used to divert calls through.

On April 17, 2005, the source code for *ToneLoc* was released.

21.1 See also

- WarVOX

21.2 References

[1] "5637 58K Users' Guide". 3COM. Retrieved 10 January 2013.

21.3 External links

- ToneLoc v1.10 source code
- ToneLoc v0.98 User Manual
- Interview with Minor Threat, ToneLoc's author, part of *BBS: The Documentary*

Chapter 22

Van Eck phreaking

Van Eck phreaking is a form of eavesdropping in which special equipment is used to pick up side-band electromagnetic emissions from electronics devices that correlate to hidden signals or data for the purpose of recreating these signals or data in order to spy on the electronic device. Side-band electromagnetic radiation emissions are present in and, with the proper equipment, can be captured from keyboards, computer displays, printers, and other electronic devices.

Van Eck phreaking of CRT displays is the process of eavesdropping on the contents of a CRT by detecting its electromagnetic emissions. It is named after Dutch computer researcher Wim van Eck, who in 1985 published the first paper on it, including proof of concept.[1] Phreaking is the process of exploiting telephone networks, used here because of its connection to eavesdropping.

Van Eck phreaking might also be used to compromise the secrecy of the votes in an election using electronic voting. This caused the Dutch government to ban the use of NewVote computer voting machines manufactured by SDU in the 2006 national elections, under the belief that ballot information might not be kept secret.[2][3] In a 2009 test of electronic voting systems in Brazil, Van Eck phreaking was used to successfully compromise ballot secrecy as a proof of concept.[4]

22.1 Basic principle

Information that drives the video display takes the form of high frequency electrical signals. These oscillating electric currents create electromagnetic radiation in the RF range. These radio emissions are correlated to the video image being displayed, so, in theory, they can be used to recover the displayed image.

22.1.1 CRTs

In a CRT the image is generated by an electron beam that sweeps back and forth across the screen. The electron beam excites the phosphor coating on the glass and causes it to glow. The strength of the beam determines the brightness of individual pixels (see CRT for a detailed description). The electric signal which drives the electron beam is amplified to hundreds of volts from TTL circuitry. This high frequency, high voltage signal creates electromagnetic radiation that has, according to Van Eck, "a remarkable resemblance to a broadcast TV signal".[1] The signal leaks out from displays and may be captured by an antenna, and once synchronization pulses are recreated and mixed in, an ordinary analog television receiver can display the result. The synchronization pulses can be recreated either through manual adjustment or by processing the signals emitted by electromagnetic coils as they deflect the CRT's electron beam back and forth.[1]

In the paper, Van Eck reports that in February 1985 a successful test of this concept was carried out with the cooperation of the BBC. Using a van filled with electronic equipment and equipped with a VHF antenna array, they were able to eavesdrop from a "large distance". There is no evidence that the BBC's TV detector vans actually used this technology, although the BBC will not reveal whether or not they are a hoax.[5]

Van Eck phreaking and protecting a CRT display from it was demonstrated on an episode of Tech TV's The Screen Savers on December 18, 2003.[6][7]

22.1.2 LCDs

In April 2004, academic research revealed that flat panel and laptop displays are also vulnerable to electromagnetic eavesdropping. The required equipment for espionage was constructed in a university lab for less than US$2000.[8]

22.1.3 Communicating using Van Eck phreaking

In January 2015, the Airhopper project from Georgia Institute of Technology, United States demonstrated (at the Ben Gurion University, Israel) the use of Van Eck Phreaking to enable a keylogger to communicate through video signal manipulation keys pressed on the keyboard of a standard PC computer, to a program running on Android cellphone with earbud radio antenna.[9][10][11]

22.1.4 Tailored Access Batteries

A tailored Access Battery is a special laptop with Van Eck Phreaking electronics and power-side band encryption cracking electronics built-into the casing of the battery in combination with a remote transmitter/receiver. This allows for quick installation and removal of spying device by simply switching the battery.[12]

22.2 Countermeasures

Countermeasures are detailed in the article on TEMPEST, the NATO's standard on spy-proofing digital equipment. One countermeasure involves shielding the equipment to minimize electromagnetic emissions. Another method, specifically for video information, scrambles the signals such that the image is perceptually undisturbed, but the emissions are harder to reverse engineer into images. Examples of this include low pass filtering fonts and randomizing the least significant bit of the video data information.

Another approach is to randomly shift the frequency of the clock used on keyboards with a custom chip containing a pseudorandom number generator (PRNG) with a long length and use an identical synchronized PRNG at the reception end to confound such attacks.

22.3 See also

- TEMPEST, a United States government standard for limiting electric or electromagnetic radiation emanations from electronic equipment

- RINT, the acronym for Radiation Intelligence, military application

- Election fraud

- Air gap (networking)

- Near sound data transfer

- SilverPush

22.4 References

[1] Van Eck, Wim (1985). "Electromagnetic Radiation from Video Display Units: An Eavesdropping Risk?" (PDF). *Computers & Security*. **4** (4): 269–286. doi:10.1016/0167-4048(85)90046-X.

[2] Dutch government scraps plans to use voting computers in 35 cities including Amsterdam (Herald tribune, 30. October 2006)

[3] Use of SDU voting computers banned during Dutch general elections (Heise, October 31. 2006)

[4] "Brazilian Breaks Secrecy of Brazil's E-Voting Machines With Van Eck Phreaking". *Slashdot*. November 21, 2009.

[5] Carter, Claire (27 September 2013). "Myth of the TV detector van?". *The Daily Telegraph*. Telegraph Media Group. Retrieved 27 September 2015.

[6] Van Eck Phreaking

[7] The Screen Savers: Dark Tip - Van Eck Phreaking

[8] Kuhn, M.G. (2004). "Electromagnetic Eavesdropping Risks of Flat-Panel Displays" (PDF). *4th Workshop on Privacy Enhancing Technologies*: 23–25.

[9] Air-gapped computers are no longer secure, TechRepublic, January 26, 2015

[10] Original Whitepaper

[11] Airhopper demonstration video, Ben Gurion University

[12] White paper, FDES institute, 1996, page 12.

22.5 External links

- Van Eck phreaking

- Van Eck phreaking Demonstration

- Tempest for Eliza is a program that uses a computer monitor to send out AM radio signals, making it possible to hear computer-generated music in a radio.

- Video eavesdropping demo at CeBIT 2006 by a Cambridge University security researcher

- eckbox – unsuccessful or abandoned attempt in spring 2004 to build an open-source Van Eck phreaking implementation

- Sniffing wireless keyboard link

- - an implementation of Van Eck phreaking using certain processor instructions on a general purpose computer

Chapter 23

War dialing

War dialing or **wardialing** is a technique of using a modem to automatically scan a list of telephone numbers, usually dialing every number in a local area code to search for computers, bulletin board systems (computer servers) and fax machines. Hackers use the resulting lists for various purposes: hobbyists for exploration, and crackers - malicious hackers who specialize in breaching computer security - for guessing user accounts (by capturing voicemail greetings), or locating modems that might provide an entry-point into computer or other electronic systems. It may also be used by security personnel, for example, to detect unauthorized devices, such as modems or faxes, on a company's telephone network.

23.1 Process

A single wardialing call would involve calling an unknown number, and waiting for one or two rings, since answering computers usually pick up on the first ring. If the phone rings twice, the modem hangs up and tries the next number. If a modem or fax machine answers, the wardialer program makes a note of the number. If a human or answering machine answers, the wardialer program hangs up. Depending on the time of day, wardialing 10,000 numbers in a given area code might annoy dozens or hundreds of people, some who attempt and fail to answer a phone in two rings, and some who succeed, only to hear the wardialing modem's carrier tone and hang up. The repeated incoming calls are especially annoying to businesses that have many consecutively numbered lines in the exchange, such as used with a Centrex telephone system.

Some newer wardialing software, such as WarVOX, does not require a modem to conduct wardialing.[1] Rather such programs can use VOIP connections, which can speed up the number of calls that a wardialer can make. Sandstorm Enterprises has a patent U.S. Patent 6,490,349 on a multi-line war dialer. ("System and Method for Scan-Dialing Telephone Numbers and Classifying Equipment Connected to Telephone Lines Associated therewith.") The patented technology is implemented in Sandstorm's PhoneSweep war dialer.

23.2 Popularity

The popular name for this technique originated in the 1983 film *WarGames*.[2] In the film, the protagonist programmed his computer to dial every telephone number in Sunnyvale, California to find other computer systems. Prior to the movie's release, this technique was known as "hammer dialing" or "demon dialing", but the film introduced the method to many, such as the members of The 414s.[3] By 1985 at least one company advertised a "War Games Autodialer" for Commodore computers.[4] Such programs became common on bulletin board systems of the time, with file names often truncated to wardial.exe and the like due to length restrictions of 8 characters on such systems. Eventually, the etymology of the name fell behind as "war dialing" gained its own currency within computing culture.[2]

The popularity of wardialing in 1980s and 1990s prompted some states to enact legislation prohibiting the use of a device to dial telephone numbers without the intent of communicating with a person.

23.3 Variants

A more recent phenomenon is wardriving, the searching for wireless networks (Wi-Fi) from a moving vehicle. Wardriving was named after wardialing, since both techniques involve brute-force searches to find computer networks. The aim of wardriving is to collect information about wireless access points (not to be confused with piggybacking).

Similar to war dialing is a port scan under TCP/IP, which "dials" every TCP port of every IP address to find out what services are available. Unlike wardialing, however, a port scan will generally not disturb a human being when it tries an IP address, regardless of whether there is a computer responding on that address or not. Related to wardriving is warchalking, the practice of drawing chalk symbols in public places to advertise the availability of wireless networks.

The term is also used today by analogy for various sorts of exhaustive brute force attack against an authentication mechanism, such as a password. While a dictionary attack might involve trying each word in a dictionary as the password, "wardialing the password" would involve trying every possible password. Password protection systems are usually designed to make this impractical, by making the process slow and/or locking out an account for minutes or hours after some low number of wrong password entries.

23.4 See also

- Autodialer
- Demon dialing
- Toneloc, a war dialer for DOS.
- Wardriving
- Warflying
- WarVOX, a war dialer using VOIP providers.
- Vishing

23.5 References

[1] Next Generation 'War-Dialing' Tool On Tap

[2] Patrick S. Ryan (Summer 2004). "War, Peace, or Stalemate: Wargames, Wardialing, Wardriving, and the Emerging Market for Hacker Ethics". Social Science Research Network. Retrieved April 2, 2008.

[3] Vollmann, Michael T (director) (2015-03-10). *The 414s: The Original Teenage Hackers*. CNN.

[4] "MegaSoft Limited". *Compute!'s Gazette* (advertisement). 1985-01. p. 167. Retrieved 6 July 2014. Check date values in: |date= (help)

23.6 External links

- 8 Bit Underground Software Archive Site dedicated to collecting 8/16-bit wardialers and other related software.
- 47 C.F.R. § 64.1200(a)(7) The 2005 revision of the TCPA appears to make wardialing a federal crime in the United States.
- 2009 article about using WarVOX for an internal network scan.

Chapter 24

WarVOX

WarVOX is a free, open-source VOIP-based war dialing tool for exploring, classifying, and auditing phone systems. WarVOX processes audio from each call by using signal processing techniques and without the need of modems.[1] WarVOX uses VoIP providers over the Internet instead of modems used by other war dialers.[2] It compares the pauses between words to identify numbers using particular voicemail systems.[3]

WarVox was merged into the Metasploit Project in August 2011. [4]

24.1 See also

- H. D. Moore
- Metasploit
- Rapid7
- ToneLoc, a war dialer for DOS.
- War dialing
- w3af

24.2 References

[1] ZDnet: Metasploit's HD Moore releases 'war dialing' tools

[2] Dark Reading: Next Generation 'War-Dialing' Tool On Tap

[3] Security Focus: War dialing gets an upgrade

[4] Dark Reading: WarVOX Gets An Overhaul; Wardialing Added To Metasploit

24.3 External links

- WarVOX official website
- The Metasploit Project Metasploit Project website, which hosts the WarVOX code

Chapter 25

Matthew Weigman

Matthew Weigman is a blind American man who used his heightened hearing ability to help him deceive telephone operators and fake various in-band phone signals. Before getting arrested at the age of 18, Weigman had used this ability to become a well known phone phreaker, memorizing phone numbers by tone and performing uncanny imitations of various phone line operators to perform pranks such as swatting on his rivals.[1]

25.1 Early life

Matthew Weigman was born and raised in East Boston. Legally blind due to optic nerve atrophy,[2] he was capable of rudimentary perception of contrast in bright light.[3]

At the age of 11, Weigman came across party lines.[2] His friends said that after only a few years, he was absorbed in these party lines to the extent that he would spend days on the phone at a time.[2] Weigman learned tricks from other party line participants, recycled from telecommunication hacker groups known as Phone Phreaks of the 1980s to gain free telephone service, frighten individuals whom he disliked and sexually harass women.[4]

25.2 First offense

At the age of 14, Weigman had already gained a considerable amount of skill and knowledge about phone hacking via party lines but had not yet used these skills to do anything illegal.[1] However, when a girl refused phone sex with him during a party line session, Weigman initiated a call to 911 with a forged Caller ID pretending to be a gunman holding her and her father at gunpoint.[5] Weigman was not indicted for his crime, marking the beginning of a *"life long obsession"* for Weigman who performed as many as 60 other forged SWAT calls prior to his indictment.[1]

25.3 Learning to hack telephones

Weigman had developed a keen sense of hearing and memory. He was able to impersonate any voice, memorize phone numbers by listening to the phone tones, and he gained the skills to understand the inner working of a phone network system by listening to the different frequencies.[1] He had the ability to mimic characters he heard on television and play songs on the piano by ear.[1]

Weigman's first experience with phreaking was by accident. He was on a party line when he discovered that by pressing the star and pound keys, he could gain access to the numbers of all the callers on the party line. He realized that he could get the number of anyone that angered him and he could harass them at their home phone number.[1] At the age of 14, Weigman learned to gain access to Verizon and AT&T by imitating an employee of the company. Weigman was known for "conning telecom employees into believing he was a colleague to gain access to unlisted numbers, the ability to shut off a rival's service or listen in on others' calls".[6]

Weigman learned to phreak phones and phone networks; consequently he was able to shut off one's phone service, dig up unlisted numbers, and listen in on conversations. Weigman also extensively employed the use of Caller ID spoofing by purchasing commercial services such as spoof cards.[1] This allowed him to hide the identity of the phone number he was using, and also to choose any number he wished to display on the caller ID on the receiving end of the call.[1]

25.4 Current incarceration

On June 26, 2009, Matthew Weigman was sentenced to 11 Years and 3 months by U.S. District Judge Barbara M.G. Lynn for convictions related to his involvement in a swatting conspiracy. He was in federal custody since being arrested in May 2008 in Boston.[5] Weigman was involved in making threats to a Verizon Security Officer, and attempted to hack into an US attorney's voice-mail system in Dallas.[5] Weigman pleaded guilty to:

"…one count of conspiracy to retaliate against a witness, victim or an informant, and one count of conspiracy to commit access device fraud and unauthorized access of a protected computer."[7]

Weigman admitted that he and his allies gained unauthorized access to telecommunication companies' sensitive information to gather personal information on certain people. He also pleaded guilty to using software to modify telecommunication devices to gain free telephone service and to cut lines of other telephone subscribers.[7]

On May 18, 2008, Weigman and others drove to the residence of the Verizon investigator who was investigating Weigman's activity, and attempted to intimidate and frighten him. Weigman had admitted that he had already placed intimidating and harassing messages and calls to the investigator.[8] When Weigman's phone line was disconnected by Verizon because the phone line he was using was illegitimate, he infiltrated the Verizon phone system and used this information to harass the employee and to gain information about the status of the investigation that Verizon was conducting.[8]

Weigman is incarcerated at the Federal Correctional Institution, Fort Dix in New Jersey. According to the Federal Bureau of Prisons, his estimated release date is May 7, 2018.[9]

25.5 See also

- Frank Abagnale
- Max Butler
- John Draper
- Phreaking

25.6 References

[1] "14-year-old blind kid, angry and alone, discovered that he possessed a superpower — one that put him in the cross hairs of the FBI". Thepeoplesvoice.org. 23 August 2009. Retrieved 15 July 2011.

[2] Morrell, Dan (February 2009). "Disconnected". Boston Magazine. Retrieved 16 July 2011.

[3] "Deceptology". Retrieved 6 August 2013.

[4] http://www.theregister.co.uk/2009/06/29/phone_phreaker_sentence/

[5] Poulsen, Kevin (29 June 2009). "Blind Hacker Sentenced to 11 Years in Prison". Wired.com. Retrieved 16 July 2011.

[6] Trahan, Jason (31 August 2008). "FBI: Teen Matthew Weigman tried to hack into Dallas' U.S. attorney's phone system". The Dallas Morning New. Archived from the original on 29 September 2011. Retrieved 16 July 2011.

[7] Wilonsky, Robert (29 June 2009). "The 19-Year-Old Blind "Little Hacker" Gets 135 Months in Federal Prison For "Swatting"". The Dallas Observer. Retrieved 17 July 2011.

[8] Jacks, James T. (29 January 2009). "Individual Pleads Guilty in Swatting Conspiracy Case". The Dallas Observer. Retrieved 16 July 2011.

[9] "Inmate Locator: Matthew Weigman". Federal Bureau of Prisons. Retrieved 29 February 2012.

25.7 External links

- Matthew Weigman Criminal Resumé (PDF)

25.8 Text and image sources, contributors, and licenses

25.8.1 Text

- **Phreaking** *Source:* https://en.wikipedia.org/wiki/Phreaking?oldid=737556959 *Contributors:* Bryan Derksen, Tarquin, Fubar Obfusco, Maury Markowitz, Sara Parks Ricker, Olivier, Citizenzero, Frecklefoot, RTC, Michael Hardy, Kwertii, Pnm, Dori, CesarB, Ahoerstemeier, Notheruser, Michael Shields, Alex756, Wfeidt, Dwo, Fry-kun, Mbstone, RickK, Ike9898, Paul Stansifer, Dysprosia, Geary, Rvolz, Furrykef, Saltine, Betterworld, Fvw, Bloodshedder, Shantavira, Denelson83, EdwinHJ, Dale Arnett, Fredrik, Greudin, Chancemill, TimothyPilgrim, Steeev, Auric, Jondel, Danceswithzerglings, Cyrius, Pengo, Falkonkirtaran, Skriptor~enwiki, Everyking, OrbitalBundle, Curps, Tieno, Beta m, Rchandra, Falcon Kirtaran, Matt Crypto, Pne, Peter Ellis, Wmahan, Luciolucolucio, Ddhix 2002, Sayeth, Hellisp, Resister, Chmod007, Chane~enwiki, Rfl, VCA, 4pq1injbok, KneeLess, Bneely, Vsmith, Smyth, Chowells, R.123, SocratesJedi, Paul August, Suriyawong, Mr. Billion, Kiand, Adrian~enwiki, Nicke Lilltroll~enwiki, Makomk, Juzeris, Larryv, Anthony Appleyard, Fwb44, Water Bottle, Stephen Turner, Seancdaug, Here, Cburnett, Anthony Ivanoff, H2g2bob, Galaxiaad, Angr, Woohookitty, Myleslong, Krille, The Wordsmith, BriskWiki, Hbdragon88, TotoBaggins, Karam.Anthony.K, Graham87, Stromcarlson, Ronnotel, Bilbo1507, BD2412, JIP, Grammarbot, Josh Parris, Koavf, Chrisp510, PinchasC, Seraphimblade, Krash, The wub, Syced, FlaBot, Latka, Nihiltres, Gary D Robson, Bmicomp, Planetneutral, Jpkotta, ColdFeet, YurikBot, Wavelength, Ailag~enwiki, Hairy Dude, Kerowren, Gaius Cornelius, Lusanaherandraton, A314268, Wiki alf, Janarius, THB, Black Ratchet, Zypres, Moe Epsilon, Voidxor, Elkman, Sir Isaac, Tawal, Deltalima, Delirium of disorder, Dkgoodman, Arthur Rubin, Sturmovik, TomHawkey, Jonathan.s.kt, MansonP, Goob, Almostc, User24, SmackBot, Elonka, Rtc, KnowledgeOfSelf, Pgk, Rrius, Dazzla, Tranced-Out, Skizzik, Saros136, Amatulic, EncMstr, SchfiftyThree, Kostmo, Hgrosser, Can't sleep, clown will eat me, Shalom Yechiel, Ianmacm, Kevlar67, Pretorious, Guroadrunner, Savetz, MKC, Rafert, RomanSpa, Othtim, Peyre, DabMachine, JmanA9, JoeBot, Highspeed, Twas Now, Dycedarg, Nczempin, Kylu, NickW557, Natas802, Lucky225, Neelix, No11akersfan, Minilek, Mr.weedle, DumbBOT, Alaibot, Wintermute314, JohnInDC, Squidward tortelini, Qwyrxian, Jedibob5, Link Spam Remover, Vaniac, Escarbot, Radimvice, Oducado, Gigi head, JAnDbot, Albany NY, Tqbf, Bongwarrior, JNW, Xb2u7Zjzc32, Leftblank, JanGB, Jim Douglas, Steven Walling, Nyttend, P.B. Pilhet, Shuini, I-baLL, MartinBot, CliffC, Jeannealcid, Jim.henderson, Rhlitonjua, Bemsor, R'n'B, KTo288, Lilac Soul, Doranchak, Piercetheorganist, Galifrag, Terabandit, Davidm617617, Peterhgregory, Black Walnut, Seanbo, VolkovBot, SupaPhreak, TXiKiBoT, Anonymous Dissident, H3xx, David Condrey, Softtest123, Pious7, Enigmaman, Haseo9999, Lamro, Edkollin, Anonymouspshreaker, Celain, Phreaka Dude, FlyingLeopard2014, Trackinfo, Jimb20, Vortalux, RMB1987, Lightmouse, Seedbot, Svick, Retractor, Tegrenath, Twinsday, ClueBot, Pressforaction, Leatherstocking, Xitit, Dgabbard, Jotag14, Draxor99, Niceguyedc, Ottava Rima, SamuelTheGhost, Tlatseg, Alexbot, Mrchris, Eeekster, Goon Noot, EutychusFr, Johnuniq, Vanished User 1004, AlanM1, Badmachine, Ost316, Asrghasrhiojadrhr, Addbot, Leszek Jańczuk, MrOllie, Mphilip1, Devinriley, Luckas-bot, Ericandrsn, Yobot, Will Decay, Synchronism, AnomieBOT, Sidfilter, Theoprakt, Xqbot, The sock that should not be, Gidoca, Multixfer, Rohitdua, Miyagawa, Tabledhote, Ace of Spades, Rkr1991, Menilek, Kgrad, Lotje, Vrenator, Tbhotch, Sideways713, RjwilmsiBot, NameIsRon, WikitanvirBot, Dewritech, Mo ainm, EyeExplore, Amilianithiantha, H3llBot, Staszek Lem, Leitz31337, Cb3684, Scientific29, Ego White Tray, ClueBot NG, Frankienoone, Widr, Calabe1992, JohnChrysostom, MusikAnimal, Jimw338, Cyberbot II, JurgenNL, SoledadKabocha, Cerabot~enwiki, Corn cheese, Electracion, IanDGunn, Phreaker007, Monkbot, Abhishekkr101, Licknooft, Dsprc, KH-1, DanielKnights, Buntee2, Matt Da Freak, TakanashiRikka25, Themikebest, Hesham Hussain, GreenC bot, Z8Moni and Anonymous: 372

- **2600 hertz** *Source:* https://en.wikipedia.org/wiki/2600_hertz?oldid=718188432 *Contributors:* Karada, Radiojon, Hajor, Jleedev, RealGrouchy, Marcika, Ferdinand Pienaar, Kar98, Smyth, Wtshymanski, Cburnett, Guthrie, H2g2bob, Arjenvr, Twthmoses, Rchamberlain, Lionel Elie Mamane, Vegaswikian, AshyRaccoon, RussBot, Deville, Sturmovik, User24, SmackBot, Sprocket, Dicklyon, Highspeed, Cornlad, Davewho2, Upholder, Jim.henderson, Mabu2, Haseo9999, Quantpole, Trackinfo, ClueBot, Bde1982, Stepheng3, XLinkBot, Lightbot, 4twenty42o, Woolfool, Adrignola, Pepper, Tbhotch, Cvs26 and Anonymous: 25

- **Mark Abene** *Source:* https://en.wikipedia.org/wiki/Mark_Abene?oldid=713973457 *Contributors:* Phr, Topbanana, Sdedeo, JustinHall, Michael Snow, DNewhall, Sam Hocevar, Chmod007, D6, Rich Farmbrough, YUL89YYZ, Xezbeth, Abelson, Night Gyr, John Vandenberg, Shenme, Adrian~enwiki, Timsheridan, Klaser, Danthemankhan, H2g2bob, Woohookitty, Myleslong, Before My Ken, Amatus, GregorB, Marudubshinki, RichardWeiss, BD2412, Josh Parris, Mayumashu, Carbonite, SchuminWeb, Flydpnkrtn, Hall Monitor, Rob T Firefly, NTBot~enwiki, Family Guy Guy, Conscious, Froth, Ospalh, Jbattersby, Deville, UltimatePyro, SmackBot, Phiberoptik, CRKingston, 6Akira7, Princemarko, Chris the speller, Pirhounix, Jjhjjh, Savidan, Volt4ire, Grimhim, Tim from Leeds, SMasters, SubSeven, Pegasus1138, Davidbspalding, Darth Maddolis, Drinibot, Neelix, Chrisdab, Blancmange, Netw1z, RobotG, Shoemaukertuvvick, LeedsKing, Bongwarrior, Grim Revenant, Instantnematode, Imecstatic, Chahax, EmxBot, Harry-, ImageRemovalBot, ClueBot, Niceguyedc, Himayat-Anjuman-i, Quercus basaseachicensis, Monobi, SchreiberBike, Miami33139, Dthomsen8, Bywater100, Addbot, Tassedethe, Lightbot, Legobot, Yobot, AnomieBOT, Materialscientist, Resident Mario, Theloufus, Mark abene, Full-date unlinking bot, Ze-dark-lord, Helpful Pixie Bot, SkateTier, Dsprc, Professornova, KasparBot and Anonymous: 63

- **BBS: The Documentary** *Source:* https://en.wikipedia.org/wiki/BBS%3A_The_Documentary?oldid=731668415 *Contributors:* ZoeB, Jscott, AnonMoos, YUL89YYZ, R.123, Bender235, Closeapple, DaveGorman, Ringbang, Myleslong, Josh Parris, RussBot, Jmchuff, SmackBot, Rtc, Unforgettableid, Ohnoitsjamie, Thumperward, D-Rock, In The Mindway, Sayden, SilkTork, Neelix, Cydebot, Lugnuts, Studerby, Shawn in Montreal, ImageRemovalBot, Trivialist, Myself248, Lightbot, Sometime-science-editor, Yobot, Polisher of Cobwebs, RayneVanDunem, BG19bot, Normosphere, Naxa and Anonymous: 14

- **BlueBEEP** *Source:* https://en.wikipedia.org/wiki/BlueBEEP?oldid=681891155 *Contributors:* MBisanz, Guthrie, Woohookitty, Srleffler, RussBot, Caerwine, Pegship, User24, Thumperward, MarshBot, Itsthemechanic, ImageRemovalBot, Yobot, Shaiku, BG19bot, Boliveirabr, Windoze96 and Anonymous: 4

- **Direct Access Test Unit** *Source:* https://en.wikipedia.org/wiki/Direct_Access_Test_Unit?oldid=635330425 *Contributors:* Klemen Kocjancic, ArnoldReinhold, Timsheridan, Kelly Martin, Bluebot, X64, Alaibot, R'n'B, Trackinfo, Dawynn, Lightbot and Anonymous: 2

- **John Draper** *Source:* https://en.wikipedia.org/wiki/John_Draper?oldid=737410116 *Contributors:* William Avery, Tuomas Toivonen, Zippy, Hephaestos, Olivier, Leandrod, Frecklefoot, Cprompt, Jpatokal, Milkfish, Michael Shields, Alex756, Camster342, Jensp~enwiki, Andrevan, RickK, Dandrake, WhisperToMe, Deselms, Furrykef, Robbot, Donreed, Cdespinosa, Perl, Frencheigh, Remy B, Mboverload, Alan Au, Golbez,

Bumm13, Chmod007, Mennonot, D6, Rfl, N328KF, Rich Farmbrough, R.123, Sourcecode, Iron Wallaby, Viriditas, 4v4l0n42, Gargaj, SidP, Voxadam, Woohookitty, Thorpe, Skyraider, Jdunck, Rchamberlain, Marudubshinki, Amorrow, Rjwilmsi, Joe Decker, Koavf, JoshuacUK, Nneonneo, FlaBot, HERMiT cRAB, Gary D Robson, Intgr, Debivort, Aalegado, Cmarie~enwiki, Moe Epsilon, Zwobot, Seigneur101, Scottfisher, Cosmotron, TransUtopian, Katpatuka, MCB, The Fish, Rex Nebular, Back ache, SOTNBot~enwiki, NiTenIchiRyu, SmackBot, Mmernex, Kellen, J4kk, Wcquidditch, McGeddon, CRKingston, Doc Strange, Pirhounix, Rlevse, Hgrosser, Veggies, Can't sleep, clown will eat me, Frap, OrphanBot, Konczewski, T-borg, Wisco, Rexmorgan, Ifrit, Will Beback, Barbalet, Ser Amantio di Nicolao, Trevor W. McKeown, Mgiganteus1, Stephlet, HelloAnnyong, Exformation, Chris55, Will314159, Paulmlieberman, Damiantgordon, Nczempin, Drinibot, Ken Gallager, DumbBOT, Uganson, JohnInDC, Xanthis, Epbr123, Blakeops, Al Lemos, Maximilian Schönherr, Born2bwire, RobotG, Waptek, Gavia immer, Dream Focus, Magioladitis, Xb2u7Zjzc32, Catgut, Bvold, Poetdancer, LorenzoB, Sneakers55, KTo288, SagaciousAWB, Lots42, Rkaufman13, Raryel, GeneralBelly, Jacktrue, Andy Dingley, Drutt, Keepfrozen, Jehorn, Macleod199, Trackinfo, TJRC, Jojalozzo, Android Mouse Bot 3, JohnSawyer, Joetrip, ClueBot, Hutcher, Hustvedt, Everlastingphelps, Alexbot, Jusdafax, Lartoven, Thingg, Sinbound, DumZiBoT, Semitransgenic, XLinkBot, Ost316, Aruhnka, ZooFari, SkyLined, Access Time, Mrsgingerbread, Addbot, Some jerk on the Internet, Zarcadia, Freqsh0, Chzz, Favonian, Aliasubik, Lightbot, Luckas-bot, TheSuave, Yobot, Twm3, Bbb23, AnomieBOT, GetMKWearMK-Fly, Materialscientist, Object404, PabloCastellano, MY MOM WONT LET ME EAT AT THE TABLE WITH A SWORD., J JMesserly, Gap9551, Wilsonchas, RibotBOT, MerlLinkBot, MandelBot, DefaultsortBot, RedBot, Sebaso, SpaceFlight89, EdoDodo, Dutchmonkey9000, Adamstrangelove, Reaper Eternal, Geoharmonics, RjwilmsiBot, ButOnMethItIs, Beyond My Ken, Balph Eubank, SaltwaterSky, Dewritech, Rahmenyah, Adrian Nyx, Heythereitsme, ClueBot NG, Danielfrost114, BG19bot, KrisAcker, Frze, Johnny Squeaky, Dunderjeep, BattyBot, ChrisGualtieri, Sailing Dutchman, VIAFbot, Rotlink, Me, Myself, and I are Here, Zenciler, YiFeiBot, Tobiastimpe, Hempmandan, Prisencolin, SpeakOnlyTheTruth404, KasparBot, Tpdwkouaa and Anonymous: 174

- **The Hacker Crackdown** *Source:* https://en.wikipedia.org/wiki/The_Hacker_Crackdown?oldid=735277999 *Contributors:* Bryan Derksen, Citizenzero, Zache, Frecklefoot, Pnm, Wwwwolf, Lquilter, Yann, Topbanana, Pmsyyz, Kjoonlee, H2g2bob, Josh Parris, Koavf, Terinjokes, RussBot, Vlad, Pegship, Fang Aili, SmackBot, Gilliam, Sdalmonte, Jerome Charles Potts, Cast, John, JesseBikman, SubSeven, Davidbspalding, Doceddi, Neelix, Skomorokh, TAnthony, Philipmac, Juliancolton, Randy Kryn, Cirt, Ottre, Addbot, Freqsh0, Lightbot, Yobot, Singularity Rider, FrescoBot, Pyxzer, CobraBot and Anonymous: 13

- **Joybubbles** *Source:* https://en.wikipedia.org/wiki/Joybubbles?oldid=736700284 *Contributors:* Deb, SimonP, Paul A, Tregoweth, Zoicon5, Moondyne, Lupo, HaeB, Gamaliel, Andycjp, DragonflySixtyseven, Bumm13, Jcm, Bneely, R.123, Bender235, Kross, Sherurcij, Guthrie, Anamanfan, Yurivict, Richard Weil, Thryduulf, Canadian Paul, Firien, GregorB, SDC, Marudubshinki, Graham87, Rjwilmsi, JanSuchy, Hall Monitor, YurikBot, Wavelength, Chanlyn, Rob T Firefly, RussBot, Cliffb, Durval, Otto Normalverbraucher, Brian Patrie, Plapsley, McGeddon, Chris the speller, Hgrosser, Vid, Frap, Konczewski, Cribcracker, Freek Verkerk, Chalub~enwiki, Blaze33541, RomanSpa, SubSeven, Cydebot, Parzi, Mister blint, Blakeops, Aempinc, Oerjan, Johnpaulb, RobotG, Tstrobaugh, Marginalia, Dmine45, Sgreddin, Themoodyblue, Jet082, Jevansen, Hqb, TravelingCat, Hluboka1, Trackinfo, Dorksgetlaid2, The BBB, Andrew8675309, Rawfasting, Mx3, Hyksos, Trivialist, PixelBot, Conical Johnson, Replysixty, Sinbound, DumZiBoT, MystBot, Kbdankbot, Addbot, Bender21435, Sillyfolkboy, Matěj Grabovský, Yobot, Omnipaedista, Surv1v4l1st, Serialjoepsycho, Pinnygold, NotWith, Sailing Dutchman, Casta947, Marc Bago, Monkbot, Blart versenwald, KasparBot and Anonymous: 57

- **Patrick K. Kroupa** *Source:* https://en.wikipedia.org/wiki/Patrick_K._Kroupa?oldid=719561939 *Contributors:* Lquilter, RedWolf, Timrollpickering, Alan Liefting, Gamaliel, Neilc, Bumm13, Neutrality, D6, Rich Farmbrough, Randomuser0101, Bender235, Adrian~enwiki, Guthrie, Galaxiaad, Bobrayner, Woohookitty, Myleslong, Easyas12c, Ferg2k, BD2412, Kbdank71, Heah, Titoxd, Tedder, RussBot, Howcheng, Deville, Eitch, That Guy, From That Show!, SmackBot, 6Akira7, Timeshifter, TrancedOut, Hmains, Rearden Metal, Chris the speller, Bluebot, Colonies Chris, Can't sleep, clown will eat me, Kristenq, OrganicChemist, Senseitaco, Beetstra, Meco, CapitalR, Davidbspalding, CmdrObot, Tanthalas39, Drinibot, ShelfSkewed, Neelix, Cydebot, Treybien, Nick Number, RobotG, Skomorokh, P L Logan, Tqbf, Wildhartlivie, Pedro, Appraiser, Aboutmovies, SmackTacular, Dj stone, Sapphic, Iaroslavvs, LarRan, ImageRemovalBot, Colonel Rozzo, Niceguyedc, Mr.Atoz, SoxBot, Ph.D. of the M.I.C., Kbdankbot, Psychotic Hitchhiker, Tassedethe, Lightbot, AnomieBOT, Sidfilter, Thermal Soldier, Novaseminary, Full-date unlinking bot, Primefac, Helpful Pixie Bot, Khazar2, I am One of Many, Valetude, SNUGGUMS, Dsprc, KasparBot and Anonymous: 24

- **Elias Ladopoulos** *Source:* https://en.wikipedia.org/wiki/Elias_Ladopoulos?oldid=726892955 *Contributors:* RadioFan, Magioladitis, LaMona, Werldwayd, Yobot, AnomieBOT, Wbm1058, BattyBot, FoCuSandLeArN, KasparBot, RMTerenzio and Russiansuperstar

- **Lucky225** *Source:* https://en.wikipedia.org/wiki/Lucky225?oldid=723067854 *Contributors:* William Avery, Jackol, Adrian~enwiki, Alai, Rjwilmsi, Alphachimp, Conscious, Closedmouth, SmackBot, Elonka, MMX, X64, Elvrum, Lucky225, I already forgot, Dylan Lake, Whammy2600, Magioladitis, VoABot II, Waacstats, WikiXan, Gwern, Nono64, Yauch, Helenabella, RogDel, JBsupreme, Yobot, Piano non troppo, Tom.Reding, RjwilmsiBot, John of Reading, Scientific29, ClueBot NG, Widr, K7L, Denniscabrams, KasparBot and Anonymous: 28

- **Phone hacking** *Source:* https://en.wikipedia.org/wiki/Phone_hacking?oldid=736161422 *Contributors:* Dreamyshade, The Anome, Lquilter, Vsmith, ZeroOne, Philip Cross, Admrboltz, GünniX, Bgwhite, Sceptre, Aeusoes1, Katieh5584, SmackBot, Xaosflux, Apeloverage, Ianmacm, Seadog365, Karenjc, Gogo Dodo, Rothorpe, Xb2u7Zjzc32, DGG, KTo288, Mdmahir, Speciate, David Condrey, Laoris, Tiptoety, Helenabella, Socrates2008, Goodone121, Addbot, Yobot, I dream of horses, Spidershadow, Traffic888, Dakisan, Onel5969, Mean as custard, Puffin, ClueBot NG, Wdchk, Widr, Helpful Pixie Bot, Wasbeer, RandomLettersForName, Nik94xx, Ddcm8991, Nova80, Skr15081997, JohnPeyton, Saectar, Monkbot, Callmeconvert, Boobsss, GabiLinko, Ansnda quadros, Bobbyisbae, Yoozay, Angelzkid212349, Jasonchua08, Sajjad.hooshi, Mohit jhanjhotiya, Tomatoestew, SwampFox221, Briankennethswain and Anonymous: 37

- **Phantom Access** *Source:* https://en.wikipedia.org/wiki/Phantom_Access?oldid=577503929 *Contributors:* Orangemike, Rich Farmbrough, Myleslong, Josh Parris, Rjwilmsi, RussBot, Gaius Cornelius, 6Akira7, TrancedOut, Drn8, Loadmaster, Neelix, R'n'B, AMbot, Mr.Atoz, Yobot, Thermal Soldier, Kithira, Jeremystalked and Anonymous: 1

- **Phone Losers of America** *Source:* https://en.wikipedia.org/wiki/Phone_Losers_of_America?oldid=735184994 *Contributors:* Timo Honkasalo, Tregoweth, Saint-Paddy, Zoicon5, Denelson83, 0x6D667061, Gzornenplatz, Wiki Wikardo, Chowbok, Zarvok, Rdsmith4, Sam Hocevar, Adashiel, Monkeyman, ElTyrant, Andy29, R.123, ESkog, Mpnolan, CanisRufus, Causa sui, Cmdrjameson, Jackliddle, Mailer diablo, El Gordo Uno, Firsfron, Henrik, Myleslong, Josh Parris, Sjakkalle, Rjwilmsi, Leoz01, Gurch, CountyKid, Rob T Firefly, RussBot, Gaius Cornelius, Arichnad, Waterguy, Tokachu, Irishguy, Countykid465, Elkman, Deltalima, Auroranorth, Soir, SmackBot, Elonka, WookieInHeat,

Delldot, Chris the speller, Lord DBK, CrookedAsterisk, Dlohcierekim's sock, Lexlex, Firetrap9254, Dreadstar, ArbeitMachtFrei, Spin359, Ardenn, Savetz, Murd0c516, Meco, Jerdobias, Ryanjunk, CmdrObot, MathewDill, Ravensfan5252, Risczero, Hitrish, Saintrain, ChrisTomkinson, Wengero, Brian Katt, Instinct, Magioladitis, VoABot II, JNW, Ling.Nut, Chris G, Ganonscrub, Logicbox, Gwern, RockMFR, Svetovid, Thomas Larsen, SJP, Starbucks95905, Oshwah, Sidepocket, TheCleanUpCrew, Ulf Abrahamsson~enwiki, StAnselm, TJRC, Winchelsea, Flyer22 Reborn, Torchwoodwho, ClueBot, Fylerfurdon, Lmkpolili, The Thing That Should Not Be, Wadiitlopu, Eeepl, Wanklop, Nonocgr, Boing! said Zebedee, Trivialist, 718 Bot, Spitfire, Killkola, Proofreader77, Nathanjamesbaker, Twinzor, Macadamiaman, Lightbot, Fraggle81, Bility, AnomieBOT, Materialscientist, Bigtruck77, LilHelpa, Midhart90, Werieth, Jamiethompsonca, ClueBot NG, Revit x man, Mythpage88, BattyBot, HollidayMasterofMystery, User0010, Kevin12xd, TheCrimsonLegacy, NigeriaNoKamisama, Daltonlux, PowerOfGamers, TheDaJakester, MrSilverizer, BlackGhost2280, JJMC89 and Anonymous: 136

- **Phreaking boxes** *Source:* https://en.wikipedia.org/wiki/Phreaking_boxes?oldid=679278466 *Contributors:* Dcandeto, Discospinster, MBisanz, Shaddack, Arthur Rubin, Chris the speller, Iridescent, NorthernThunder, Javawizard, Mrchris, Alex.blackbit, Lightbot, Omnipaedista, Cnwilliams, Gamewizard71, Mdann52, Khazar2, K7L, Mogism, Phreaker007, Gab4gab and Anonymous: 5

- **Plover-NET** *Source:* https://en.wikipedia.org/wiki/Plover-NET?oldid=734773028 *Contributors:* WordsEchoThus, SubSeven, Magioladitis, Satani, David Condrey, Carbinga, Dsprc, Peterbrooke581, CLCStudent, GeneralJohnsonJameson and Anonymous: 1

- **The Secret History of Hacking** *Source:* https://en.wikipedia.org/wiki/The_Secret_History_of_Hacking?oldid=646990629 *Contributors:* Hack, Fram, Ironholds, AnomieBOT, FrescoBot, Fortdj33, Jonkerz, ClueBot NG, Sailing Dutchman and Anonymous: 2

- **StankDawg** *Source:* https://en.wikipedia.org/wiki/StankDawg?oldid=725282934 *Contributors:* Tregoweth, Jscott, Bearcat, Kevin B12, Trevor MacInnis, Timothy Campbell, Rich Farmbrough, Giraffedata, Timsheridan, Mike.mihaylov, H2g2bob, Stemonitis, Woohookitty, Stromcarlson, Rjwilmsi, Jehochman, Syced, Ian Pitchford, Thebracket, Hall Monitor, RussBot, Rsrikanth05, Haemo, Sarah, SmackBot, Elonka, Anastrophe, Moe Aboulkheir, Chris the speller, Ned Scott, Can't sleep, clown will eat me, Othtim, ShelfSkewed, Neelix, Cydebot, RFerreira, RobotG, Ministry of Truth, KDerrida, OhanaUnited, Tqbf, Waacstats, MetsBot, Patch Cable, R'n'B, PDFbot, CutOffTies, Bad Monk3y, CounterVandalismBot, Coccyx Bloccyx, Tothwolf, Tassedethe, OlEnglish, LilHelpa, Jonesey95, RjwilmsiBot, John of Reading, Edw400, Codename Lisa, Hmainsbot1, Monkbot, KasparBot and Anonymous: 38

- **Dennis Terry** *Source:* https://en.wikipedia.org/wiki/Dennis_Terry?oldid=718672884 *Contributors:* Moondyne, Bgwhite, Cydebot, Waacstats, Trackinfo, Arbor to SJ, Jojalozzo, Niceguyedc, Yobot, AnomieBOT, Sailing Dutchman, KasparBot and Anonymous: 3

- **ToneLoc** *Source:* https://en.wikipedia.org/wiki/ToneLoc?oldid=735156607 *Contributors:* Geoffrey~enwiki, Schneelocke, Furrykef, Cyrius, Lucky 6.9, R.123, Grutness, Tabor, SidP, BDD, Myleslong, RussBot, Rsrikanth05, BOT-Superzerocool, Pegship, Addbot, Lightbot, Yobot, Pradameinhoff, BlackGhost2280, Bender the Bot and Anonymous: 10

- **Van Eck phreaking** *Source:* https://en.wikipedia.org/wiki/Van_Eck_phreaking?oldid=734306607 *Contributors:* Damian Yerrick, HelgeStenstrom, Dominus, Tregoweth, Stw, Julesd, Bevo, Lesonyrra, Bkell, Graeme Bartlett, Mako098765, DragonflySixtyseven, Patilkedar, Omassey, Xezbeth, Antaeus Feldspar, Kghose, John Vandenberg, Rebroad, Danhash, Rshin, Gene Nygaard, Falcorian, Oliphaunt, Jacobolus, MiG, Koavf, SeanMack, KaiMartin, Riki, Todd Vierling, RussBot, Juansmith, Hydrargyrum, Shinmawa, Speedevil, GrafZahl, SmackBot, Fulldecent, Antifumo, Commander Keane bot, Bluebot, Imaginaryoctopus, Frap, Jmnbatista, Cybercobra, Bogsat, Michael Rogers, Cdills, Shdwsclan, Bc.rox.all, Cowicide, Pimlottc, Cxw, Maester mensch, Lmcelhiney, ShelfSkewed, Boemanneke, Mbell, Ekpyyrotic, Aifesteves, Rehnn83, Btrotter, Seeve, Lklundin, PBot, Jasper Janssen, Electiontechnology, R'n'B, PCock, KiwiBiggles, Taintain, Squids and Chips, Burpen, Wingedsubmariner, Spinningspark, MotherForker, Launchpad 72, Niceguyedc, Jomsborg, Dhugot, Addbot, DOI bot, Lightbot, Yobot, Plazmatyk, Kav2k, Kylexysf, Kylexysf1, Philleb, Bzzz, Citation bot 1, Wingman4l7, DanReiss, KarenLFranklin, Shaztastic, Stub Mandrel, 24ghz, Michielap, Smirkhuts, FtLauderGuy and Anonymous: 62

- **War dialing** *Source:* https://en.wikipedia.org/wiki/War_dialing?oldid=715397546 *Contributors:* Fubar Obfusco, Hephaestos, Frecklefoot, Liftarn, Tannin, Wwwwolf, Theresa knott, Arteitle, AC, Furrykef, Omegatron, Donreed, Altenmann, Psychonaut, Greudin, Dave6, Jonabbey, J3ff, Jtmendes, FiP, Ylee, Klipper~enwiki, Dowcet, Woohookitty, Simsong, Dpaking, Graham87, Josh Parris, FlaBot, Gurch, Family Guy Guy, RussBot, Kvn8907, Kedwiki, Kkmurray, Alecmconroy, KnightRider~enwiki, SmackBot, Melchoir, McGeddon, Shabda, Canthusus, Thenickdude, NYKevin, Kanabekobaton, JonHarder, Steve Pucci, Kukini, JzG, Zaphraud, Stratadrake, Dav Brav, OnBeyondZebrax, Mellery, JohnCD, Orderinchaos, ShelfSkewed, Neelix, No1lakersfan, Equendil, Rifleman 82, Benjiboi, Thijs!bot, Tgeairn, J.delanoy, USN1977, Yauch, Dispenser, White 720, Vranak, TXiKiBoT, Johnred32, Bodybagger, Centerone, K1ng l0v3, Android Mouse Bot 3, Riking8, John Nevard, Replysixty, Aitias, InternetMeme, Addbot, Protonk, LaaknorBot, OlEnglish, Fiftyquid, Marshall Williams2, AnomieBOT, Unara, Pradameinhoff, Erik9, DAVilla, Limited Atonement, Coloursoftherainbow, ClueBot NG, Helpful Pixie Bot, BG19bot, GSS-1987 and Anonymous: 41

- **WarVOX** *Source:* https://en.wikipedia.org/wiki/WarVOX?oldid=734210469 *Contributors:* Cydebot, Goltz20707, DanielPharos, Yobot, Pradameinhoff, Nameless23, VernoWhitney, VWBot and Anonymous: 1

- **Matthew Weigman** *Source:* https://en.wikipedia.org/wiki/Matthew_Weigman?oldid=736673619 *Contributors:* Bearcat, Moondyne, Esrogs, Acsenray, Mandarax, Graham87, Rwalker, Zyxw, Zeroality, Mckraz, Crunchy Numbers, Acmilan10italia, Ng.j, JasonAQuest, Yobot, AnomieBOT, Seligne, DarwinSurvivor, Lotje, GoingBatty, Doelleri, Smittyboyfloyd, AdventurousSquirrel, ASCIIn2Bme, 220 of Borg, Cyberbot II, Agn106, Dough34, KasparBot and Anonymous: 10

25.8.2 Images

- **File:2600_Hz.ogg** *Source:* https://upload.wikimedia.org/wikipedia/commons/f/fe/2600_Hz.ogg *License:* Public domain *Contributors:* Own work *Original artist:* H2g2bob

- **File:Ambox_important.svg** *Source:* https://upload.wikimedia.org/wikipedia/commons/b/b4/Ambox_important.svg *License:* Public domain *Contributors:* Own work, based off of Image:Ambox scales.svg *Original artist:* Dsmurat (talk · contribs)

- **File:Blue_Box_at_the_Powerhouse_Museum.jpg** *Source:* https://upload.wikimedia.org/wikipedia/commons/0/0d/Blue_Box_at_the_Powerhouse_Museum.jpg *License:* CC BY-SA 4.0 *Contributors:* Own work *Original artist:* Maksym Kozlenko

25.8. TEXT AND IMAGE SOURCES, CONTRIBUTORS, AND LICENSES

- **File:Bluebeepuser.jpg** *Source:* https://upload.wikimedia.org/wikipedia/commons/f/f0/Bluebeepuser.jpg *License:* CC BY-SA 3.0 *Contributors:* Own work *Original artist:* Boliveirabr
- **File:Cap'n_Crunch,_Spielzeugpfeife_(2600_Hz).jpg** *Source:* https://upload.wikimedia.org/wikipedia/commons/d/da/Cap%E2%80%99n_Crunch%2C_Spielzeugpfeife_%282600_Hz%29.jpg *License:* CC BY-SA 4.0 *Contributors:* Own work *Original artist:* © 1971markus
- **File:Carl_Spitzweg_021-detail.jpg** *Source:* https://upload.wikimedia.org/wikipedia/commons/8/81/Carl_Spitzweg_021-detail.jpg *License:* Public domain *Contributors:* Diese Datei: File:Carl Spitzweg 021.jpg
  Original artist: Carl Spitzweg
- **File:Commons-logo.svg** *Source:* https://upload.wikimedia.org/wikipedia/en/4/4a/Commons-logo.svg *License:* CC-BY-SA-3.0 *Contributors:* ? *Original artist:* ?
- **File:Crunch-hackers.jpg** *Source:* https://upload.wikimedia.org/wikipedia/commons/5/50/Crunch-hackers.jpg *License:* CC BY-SA 4.0 *Contributors:* Own work *Original artist:* Aliasubik
- **File:Edit-clear.svg** *Source:* https://upload.wikimedia.org/wikipedia/en/f/f2/Edit-clear.svg *License:* Public domain *Contributors:* The *Tango! Desktop Project*. *Original artist:*
  The people from the Tango! project. And according to the meta-data in the file, specifically: "Andreas Nilsson, and Jakub Steiner (although minimally)."
- **File:Flag_of_the_United_States.svg** *Source:* https://upload.wikimedia.org/wikipedia/en/a/a4/Flag_of_the_United_States.svg *License:* PD *Contributors:* ? *Original artist:* ?
- **File:Folder_Hexagonal_Icon.svg** *Source:* https://upload.wikimedia.org/wikipedia/en/4/48/Folder_Hexagonal_Icon.svg *License:* Cc-by-sa-3.0 *Contributors:* ? *Original artist:* ?
- **File:Free_and_open-source_software_logo_(2009).svg** *Source:* https://upload.wikimedia.org/wikipedia/commons/3/31/Free_and_open-source_software_logo_%282009%29.svg *License:* Public domain *Contributors:* FOSS Logo.svg *Original artist:* Free Software Portal Logo.svg (FOSS Logo.svg): ViperSnake151
- **File:Gnome-mime-sound-openclipart.svg** *Source:* https://upload.wikimedia.org/wikipedia/commons/8/87/Gnome-mime-sound-openclipart.svg *License:* Public domain *Contributors:* Own work. Based on File:Gnome-mime-audio-openclipart.svg, which is public domain. *Original artist:* User:Eubulides
- **File:John_Draper,_Lee_Felsenstein,_Roger_Melen_(2013).jpg** *Source:* https://upload.wikimedia.org/wikipedia/commons/6/63/John_Draper%2C_Lee_Felsenstein%2C_Roger_Melen_%282013%29.jpg *License:* CC BY-SA 3.0 *Contributors:* Own work *Original artist:* Cromemco
- **File:MarkAbene1.jpg** *Source:* https://upload.wikimedia.org/wikipedia/en/1/16/MarkAbene1.jpg *License:* CC-BY-SA-3.0 *Contributors:* ? *Original artist:* ?
- **File:Motorola_L7.jpg** *Source:* https://upload.wikimedia.org/wikipedia/commons/3/3d/Motorola_L7.jpg *License:* CC BY-SA 3.0 *Contributors:* Own work (Original text: *I (Kristoferb (talk)) created this work entirely by myself.*) *Original artist:* Kristoferb (talk)
- **File:Paccess1.jpg** *Source:* https://upload.wikimedia.org/wikipedia/en/f/f1/Paccess1.jpg *License:* Fair use *Contributors:*
  The logo may be obtained from Phantom Access.
  Original artist: ?
- **File:Phone_Losers_of_America_at_the_Fifth_HOPE.jpg** *Source:* https://upload.wikimedia.org/wikipedia/commons/6/6f/Phone_Losers_of_America_at_the_Fifth_HOPE.jpg *License:* CC BY-SA 3.0 *Contributors:* Own work *Original artist:* Murd0c516
- **File:Phone_Losers_of_America_logo.svg** *Source:* https://upload.wikimedia.org/wikipedia/en/e/e6/Phone_Losers_of_America_logo.svg *License:* Fair use *Contributors:*
  http://phonelosers.org *Original artist:* ?
- **File:Phone_icon_rotated.svg** *Source:* https://upload.wikimedia.org/wikipedia/commons/d/df/Phone_icon_rotated.svg *License:* Public domain *Contributors:* Originally uploaded on en.wikipedia *Original artist:* Originally uploaded by Beao (Transferred by varnent)
- **File:Question_book-new.svg** *Source:* https://upload.wikimedia.org/wikipedia/en/9/99/Question_book-new.svg *License:* Cc-by-sa-3.0 *Contributors:*
  Created from scratch in Adobe Illustrator. Based on Image:Question book.png created by User:Equazcion *Original artist:* Tkgd2007
- **File:Radio_svg_icon.svg** *Source:* https://upload.wikimedia.org/wikipedia/commons/9/9f/Radio_svg_icon.svg *License:* CC0 *Contributors:* Own work *Original artist:* Timothy King
- **File:Synaptic.png** *Source:* https://upload.wikimedia.org/wikipedia/commons/0/05/Synaptic.png *License:* GPL *Contributors:* [1] *Original artist:* en:User:Burgundavia
- **File:Text_document_with_red_question_mark.svg** *Source:* https://upload.wikimedia.org/wikipedia/commons/a/a4/Text_document_with_red_question_mark.svg *License:* Public domain *Contributors:* Created by bdesham with Inkscape; based upon Text-x-generic.svg from the Tango project. *Original artist:* Benjamin D. Esham (bdesham)
- **File:Unbalanced_scales.svg** *Source:* https://upload.wikimedia.org/wikipedia/commons/f/fe/Unbalanced_scales.svg *License:* Public domain *Contributors:* ? *Original artist:* ?
- **File:Unitedphonelosers.gif** *Source:* https://upload.wikimedia.org/wikipedia/commons/a/ac/Unitedphonelosers.gif *License:* Public domain *Contributors:* Transferred from en.wikipedia to Commons by Explicit using CommonsHelper. *Original artist:* Jerdobias at English Wikipedia
- **File:Video-x-generic.svg** *Source:* https://upload.wikimedia.org/wikipedia/en/e/e7/Video-x-generic.svg *License:* Public domain *Contributors:* ? *Original artist:* ?

25.8.3 Content license

- Creative Commons Attribution-Share Alike 3.0

www.ingramcontent.com/pod-product-compliance
Lightning Source LLC
Chambersburg PA
CBHW080625190526
45169CB00009B/3286